NOTFINITY PROCESS

NOTFINITY PROCESS

MICROCOSMS-IN-MOTION

GEORGE S. COYNE

outskirts
press

Notfinity Process
Microcosms-in-Motion
All Rights Reserved.
Copyright © 2017 George S. Coyne
V2.0

The opinions expressed in this manuscript are solely the opinions of the author and do not represent the opinions or thoughts of the publisher. The author has represented and warranted full ownership and/or legal right to publish all the materials in this book.

This book may not be reproduced, transmitted, or stored in whole or in part by any means, including graphic, electronic, or mechanical without the express written consent of the publisher except in the case of brief quotations embodied in critical articles and reviews.

Outskirts Press, Inc.
http://www.outskirtspress.com

ISBN: 978-1-4787-7789-2

Cover Photo © 2017 thinkstockphotos.com. All rights reserved - used with permission.

Outskirts Press and the "OP" logo are trademarks belonging to Outskirts Press, Inc.

PRINTED IN THE UNITED

Grateful Acknowledgments

I greatly appreciate the work of my reviewer, Glenn Borchardt, Ph.D., director of the Progressive Science Institute, and author of several important books on theoretical physics and cosmology. His latest book, *Infinite Universe Theory* comes out in the second half of 2017. Glenn was very generous with his time in communicating with me about *Notfinity Process*. He provided many valuable comments, editorial suggestions and help in clarifications. In addition, I thank him for his kind permission to quote freely from his published material. I discuss many of his ideas extensively in several parts of this book.

The other reviewer was Steven Bryant, author of *Disruptive*, who had many useful suggestions and comments that helped in explaining some of the ideas in this book. I thank him for permission to use his English language translation of Einstein's spherical wave proof in Chapter 6. I highlight some of his most significant work, especially his critique of relativity theory.

Through my association with Steve, I was able to obtain the services of his excellent editor, Grant John Dexter, who worked on Steve's revolutionary physics book, *Disruptive* (2016). I am very grateful for the copyediting Grant did on the manuscript, which produced many improvements.

I appreciate physicist Francis David Peat giving his permission to use e-mail letters that he sent to me. I have included two of these. I was greatly saddened by his death on June 6, 2017. David was always exceedingly kind and encouraging to me in my writing. He was the first person to receive an early version of the Notfinity Process manuscript in the summer of 2016, although he decided to wait to read the finished copy, which was too late for him. The last words that he wrote to me in response to my updating him on Notfinity Process was: "Thanks for the news, Good work."

Many thanks to quantum physicist Basil Hiley for allowing me to include a letter he sent to me in August 2014 and another one on June 9, 2017. Additionally I am grateful for his clear explanation of the quantum potential in his letter to me on August 3, 2017 from which I have quoted him. It is a\rare privilege to be tutored by the world's foremost expert on the Bohm approach to quantum mechanics.

Thanks to my friend Duncan W. Shaw, retired Justice and author of several peer-reviewed physics papers, for our fruitful and intriguing conversations on theoretical physics. I have included many of his theories in this book.

Thanks to physicist and consciousness theorist Bruce Nappi of the A_3 Society for reviewing the manuscript. His comments were very useful in clarifying ideas in this book.

Many thanks to Nick Percival, who gave permission to include "An Open Letter to the Physics Community The Twin Paradox." We are both members of the board of the John Chappell Natural; Philosophy Society.

I am grateful to Stephen Puetz for permitting me to use an extended quote from *Universal Cycle Theory*.

Thanks to Ray Brooks, a shakuhachi musician, and author of *The Shadow that Seeks the Sun*, for promoting this book.

Without the positive responses from many people who were receptive to the ideas discussed in Notfinity Process, this book may not have been written. I thank Aziz Baloch for being an early supporter of this project and Doug Protz for having recommended that I share my views with a wider audience. Additional thanks to Barry Coyne for his support throughout this undertaking.

.

Dedication

In memoriam of my parents, who were tremendous examples of humanity at its finest.

In memory of the philosopher/theoretical physicist David Bohm, who contributed to the development of science and a more holistic way of perceiving ourselves as indivisible from the Universe.

In memory of Francis David Peat, who was a good friend. As a humanitarian, a holistic physicist and creator of "gentle action," his life and work contributed profoundly to making the world a better place.

In memory of Halton C. Arp, an astronomer who was a modern day Galileo, inspiring many scientists, who value discovery ahead of personal ambition.

Table of Contents

Grateful Acknowledgments 5

Dedication 7

Image 17

Meaning of Acronyms...18

The Progressive Science Institute..20

Preface 21

Part One   Assumptions..25

Chapter I..26

Important Concepts 26
   Focus and Theme 26
   The Ten Assumptions of Science Comparisons 36
      1. Materialism 37
      2. Causality 37
      3. Uncertainty... 37
      4. Inseparability 38
      5. Conservation.. 38
      6. Complementarity.. 38
      7. Irreversibility... 39
      8. Infinity 40
      9. Relativism 40
      10. Interconnection 40

Part Two Abstractions 43

Chapter 2 44

Characteristics of Abstractions 44
Understanding their Meaning 44
Abstract versus concrete objects 45
Features of Abstractions 46
Abstractions used in Professions 53

Chapter 3 56

Appropriate Role of Math in Physics 56
Views from Scientists 56
Language of Math 58
Limitations of Math in Theories 59
Math is not the Essence of Reality 61

Part Three Invalid Abstractions 65

Chapter 4 66

Big Bang Theory Falsified 66
Dozens of BBT Problems 66
Preposterous Indeterminism 66
Attraction 69
Curved Space 70
Chaos and Heat Death 70
Cosmic Expansion: Gravitational Redshift 71
Redshift, Dark Matter and Molecular Hydrogen 72
Non-expansion Velocity Causes of Redshift 74
Aether as cause of Redshift 74
Explanation of CMB 75
Cosmic Inflation Concept and Dark Matter 76
Universe is Not Expanding 79
More Evidence for No Expansion 82
Tired Light Theory (TLT) 84
Big Bang Singularity 84
Hannes Alfvén 85
The Horizon or Isotropy Problem 86
Monopole Problem 88

Table of Contents

Debate on whether BBT Violates Conservation Law 88
Doppler effect and Redshift 93
Globular Clusters and Superclusters: Age Problem 94
Age of Quasi-Stellar Objects (QSO) or Quasars 95
Quasars: Younger than Claimed 96
Violates Second Law of Thermodynamics 96
Uniform Absorption Lines 96
The Galaxy Formation Problem 96
Galaxies Developed Too Early 96
CMB Temperature 97
BBT Stopped Predicting 98
Redshift Quantization 99
Contradictions in N-body Simulations 99
Virgo's Streaming is Not Caused by an Attractor 100
Microwave Radiation Fluctuations Data 100
Contradictory Densities Problem 101
Entropic Cosmic View 101
Dark Flow Significance 102
Interpretation of Dark Flow 102
Matter Concentration: Hundreds of Billions of Years 103
BBT Needs Dark Matter 103
BBT Requires Inflation and Dark Energy 104
Matter to Antimatter Ratio Problem 105
Too Many Faint Blue Galaxies 105
Parameters for Elements Abundances 106
Quasar Redshift Brightness 106
Changing the Facts to fit BBT 106
Brightest Galaxy Clusters (BGCs) 107
Constancy of Surface Brightness 108
HUDF and GALEX Data Conflicts with BBT 108
Quasars' Faraday Effect 108
Large Structures 1.65-trillion-years old 109
Cosmos is not Homogenous and Isotropic 109
Too Few Galaxies Change from Elliptical 110
Microwave Radiation Fluctuations 111
Maximum Energy Levels Exceeded 111
Temperature of the Medium between Galaxies 111
CMB Alignment with Local Supercluster 111
Impossible Mature Galaxies in Early Universe 112
23 Billion-year-old Galaxies Contradict BBT 112

Increasing Variety of Specifications Needed ... 113
CMB is Weak Evidence ... 114
Conclusion: BBT is Invalid ... 116
Research Funds Only Go to BBT Proponents ... 117

Chapter 5 ... **125**

Absurd QMT, SRT and GRT ... **125**
Copenhagen Interpretation (CI) ... *125*
Space without Matter Abstraction ... *129*
Impossible Motion without Matter abstraction ... *129*
Abstraction of Energy ... *130*
Petroleum as Energy Abstraction ... *132*
Inaccurate Interpretations of Energy and Force ... *133*
Wave-Particle Duality ... *134*
Photons in a Vacuum ... *134*
Negative Mass ... *134*
Gravity is not Distortion of Space ... *135*
Gravity, Not Mass Increases with Velocity ... *136*
Physical Lorentz Relativity versus Observer-Centric SRT ... *140*
How SRT Derives the "Correct Result" ... *141*
SRT's Problem ... *142*
Open Letter to the Physics Community: The Twin Paradox ... *143*
Implications for Cosmology and Physics ... *152*
Establishment's Intolerance of Dissenting Views ... *152*

Part Four   Theories from Observations ... **157**

Chapter 6 ... **158**

Modern Mechanics Replaces Relativity ... **158**
General Relativity Invalidated ... *160*
Einstein's Errors ... *162*
Motion in MMT ... *164*
Ives-Stillwell ... *165*
Bryant's Reassessment of Michelson-Morley Experiment ... *166*
Speed of Light and Fine-Structure Constant ... *167*
Electromagnetic Force ... *169*
References ... *169*

Table of Contents

Chapter 7 172

QMT vs. Alternatives 172
   QMT Contradictions 172
   Matter 173
   QMT Duality of Light Concept 174
   Superposition and Probability Waves 175
   QMT Contradictions: Interpretation of Double-slit Experiment 175
   Communication from Physicist Basil Hiley 178
   Steven Bryant's MMT Interpretation 180
   Schrödinger's cat Thought Experiment 181
   Misinterpreting Entanglement Phenomena 182
   Bohm and Hiley's Entanglement Explanation 186
   Threshold Model 186
   Shaw's Elegant Solution to Entanglement 187
   Ridiculous Declaration that Universe is Impossible 189

Chapter 8 194

Theories without Contradictions 194
   Matter and Motion are Abstractions 194
   Microcosms within Macrocosms 196
   Motion Required For Microcosms 197
   Understanding the Heisenberg Uncertainty Principle 200
   Limitless Microcosms in motion 201
   Notfinity and David Peat Letter 202
   Meaning of Time as a verb 203
   Language that conforms to Reality 210
   David Peat's Letter on Bohm's Rheomode 210
   Quantized Space 212
   How Specific Motion Occurs in Infinite Space 213
   Motion, Time and Change 216

Chapter 9 219

Physics for the 21st Century and Beyond 219
   What is Dark Matter? 219
   X-ray Emission Line not dark matter 219
   Alternative Explanations for Dark Matter 220
   Maxwell's Aether Theory 220

Polarization of Aether 222
Aether Theory Implications 222
Modern Mechanics Theory 225

Chapter 10 229

Infinite Universe Theory (IUT) 229

Key Concepts 229
IUT Summary and Comparing Assumptions of Two Theories 232

Chapter 11 235

Universal Cycle Theory 235
Matter's Cyclical Movement 235
NGT Emphasizes Assumptions 4, 5, 6, 8, 9 and 10 237
Differences from Classical Mechanics 238
Vortex Formation 238
Key Features of NGT 240
Newton's GPG concept 240
Light Velocity a Function of Aether Density 244
Total-Mass Equation 244
Gravitation is a Local Phenomenon 245

Chapter 12 250

David Bohm's Approach 250
Pilot Waves 250
Causality and Chance 250
The Undivided Universe 254
Delayed-Choice Experiments and Bohm Approach 256

Chapter 13 260

Shaw's Aether Gravity Model 260
Introducing Shaw's Theory 260
Flowing Aether 260
Aether Pressure Differential 262
Aether Experimental Evidence 263
Summation and Conclusions 264

Table of Contents

Part Five   Consciousness Abstraction 267

Chapter 14 268

Functioning Brain is Consciousness 268
- *Neurological Consciousness Study* 268
- *Process of Inputs/Outputs and Consciousness* 269
- *Dualistic Confusion* 271
- *Two Minds Hypothesis* 273
- *Consciousness equates to Functioning Brain* 274
- *Triune Brain Theory* 274
- *Global Workspace Theory (GWT)* 275
- *Chunking Process* 277
- *Question about Consciousness and the Brain* 278
- *Attention and Consciousness* 280
- *"Free Will" Concept* 281
- *Negative Implications for Free Will from Brain Studies* 283
- *Consciousness as Byproduct of Entropy* 285
- *The "Observer" Illusion* 286
- *Philosophy of Mind and Predictive Processing* 287
- *Concept of Self-awareness* 288
- *Antonio Damasio's Model* 289
- *Functional Explanation of Easy Problem* 292
- *Hard Problem of Consciousness and its Solution* 293
- *Evolutionary Benefit of Advanced Mental Processes* 297
- *Summary of Discoveries of Neuroscience* 298

HOR Theories 298
- *HOR Solution to Hard Problem* 300
- *Characteristics of Higher-order Thought Theory* 300
- *Wide-Intrinsicality View (WIV)* 300
- *HOT Theory* 301
- *Higher-order Perception Theory* 302
- *Evaluation* 302
- *Computational and Representational Theories of Mind* 304
- *Self as Collection of Programs* 304

Chapter 15 311

Summary and Implications 311

Definitions **315**

Bibliography **325**
   Books *325*
   Papers and Articles *329*

Index **357**

Images

Image 1 The Hubble Ultra Deep Field 2012   113
Image 2 The magnificent starburst galaxy Messier 82.   118
Image 3 Einstein's Spherical Wave Proof.   159
Image 4 Illustration of Ives-Stillwell Experiment.   165
Image 5 Illustration Double-slit physics experiment.   175
Image 6 Illustration of Two slits illuminated by a plane.   176
Image 7 The Whirlpool Galaxy M51 (NGC 5194)
            and companion ga*laxy NGC 5195*.   240
Image 8 Newton on Gravitational Pressure Gradient.   242

Meaning of Acronyms

ACT	Atacama Cosmology Telescope
BBT	Big Bang Theory
BGC	Brightest Cluster Galaxies
BI	Bohm interpretation
BICEP	Background Imaging of Cosmic Extragalactic Polarization
CDM	Cold Dark Matter
CI	Copenhagen interpretation
CMB	Cosmic Microwave Background
COBE	Cosmic Background Explorer
CP	Cosmological Principle
EMR	Electromagnetic Radiation
EPR	Einstein, Podolsky, Rosen
FRW	Friedman-Robertson-Walker
HD	Hard Determinism
HST	Hubble
HUDF	Hubble Ultra Deep Field
GALEX	Galaxy Evolution Explorer
GPG	Gravitational Pressure Gradient
GRT	General Relativity Theory
GUT	Grand Unified Theory
GWT	Global Workspace Theory
H_2	Molecular Hydrogen
HOR	Higher- Order Representation
HOT	Higher- Order Thought
LIGO	Laser Interferometer Gravitational-Wave Observatory
IUT	Infinite Universe Theory
MOND	Modified Newtonian Dynamics

Mpc	Mega-parsec
m/s	Meters per second
MMT	Modern Mechanics Theory
MPA	Max Planck Institute for Astrophysics
NASA	National Aeronautics and Space Administration
NGT	Neomechanical Gravitational Theory
P=mv	Momentum equals mass times volume
PDE	Partial Differential Equation
SLT	Second Law of Thermodynamics
SPT	South Pole Telescope
UD	Univironmental Determinism
UCT	Universal Cycle Theory
UFT	Unified Field Theory
QE	Quantum Entanglement
QFT	Quantum Field Theory
QMT	Quantum Mechanics Theory
QP	Quantum Potential
QSO	Quasi-Stellar Objects
TLT	Tired Light Theory
TOME	Theory of Miraculous Events
TOE	Theory of Everything
UFT	Unified Field Theory
UTG	Univironmental Theory of Gravitation
WIV	Wide-Intrinsicality View
WMAP	Wilkinson Microwave Anisotropy Probe
Z	Redshift
ZPE	Zero-point energy.

The Progressive Science Institute

The Progressive Science Institute (PSI) sponsors education and research on scientific philosophy that is free of religious and political influence. It argues against cosmogony models, such as the Big Bang where the universe erupts from a void, and those stipulating greater than three dimensions.[1] PSI has been very supportive throughout the entire process of producing this book, with many very useful suggestions, including editorial input.

Glenn Borchardt, Ph.D. Director of the Progressive Science Institute (PSI) and author of several scientific books wrote the following introduction on March 2, 2016 on the Scientific Worldview website as after appointing me to be head of the Vancouver PSI:[2]

> "George came to our attention through his insightful analyses of physics and cosmology based on the Ten Assumptions of Science. We suspect that he could recite them verbatim by now.
>
> His comments and guest blogs always were right on. George's story below is strikingly similar to that of other PSI members: an inquiring mind from an early age and not one to accept paradox and contradiction as the natural state of the universe. George has agreed to head the Vancouver Regional Office of PSI."

---

[1] Progressive Science Institute, *Welcome to Scientific Philo*sophy.
[2] Borchardt, G., *The Scientific Worldview.*

Preface

My interest in science began in late 1956 at the age of four when I was attempting to conceive of the Earth in the universe. At 11 years of age I began wondering about what the word "time" actually meant. I did an experiment to try to find out the nature of time. I wanted to know whether it was possible to make an experiential connection between that moment in time and myself in 20 years. I still recall what I was thinking at the time.

Since my early teens, I have been exploring the wonders of the ways that we connect to our environment. This has taken the form of my formal academic pursuits in psychology and my explorations in philosophy, cosmology and theoretical physics.

By age 15, I no longer accepted that time was a flowing thing. At age 16 in an essay on "time," I stated that I did not know what it was and that although ideas about it will change, what the word refers to will be the same. By the time I was 20, I was sure that accepted definitions for time were complete nonsense and what they described did not exist.

I next began seeking definitions for matter, energy, and space. I was unable to find any that made sense. It appeared that scientists had no idea about how these concepts represented any particular thing or occurrence in the universe. They seemed to be using their professional power validate their views on these topics. In my early 20s, I began reading about quantum mechanics (QMT) and relativity physics. Most of the QMT books were on the Copenhagen interpretation (CI). I never saw how this explanation was remotely possible. One of the biggest disagreements I had was with the concept that the universe cannot exist unless an observation occurs, or one makes a measurement. I explain this further along with other reasons for rejecting CI in a section of Chapter 5 on the Copenhagen interpretation.

By 1981, I had become interested in David Bohm's interpretation of quantum mechanics. I was impressed with *The Undivided Universe*, co-written with Basil Hiley in 1993, because it did not contain impossible concepts, contradictions, or paradoxes, such as duality. However, I did not agree that one could alternate between holding deterministic and indeterministic worldviews depending on the situation.

On August 1, 2014, I began corresponding with quantum physicist F. David Peat, the director of The Pari Center for New Learning. Later that month I contacted. Basil Hiley, who co-wrote The Undivided Universe with David Bohm. The Majorana Prize Committee chose him as the Best Person in Physics in 2012 for his "fundamental contribution to Theoretical Physics," and "his paramount importance as natural philosopher, his critical and open minded attitude towards the role of science in contemporary culture. Dr. Hiley represents what is best in scientists." He replied quickly and positively to my letter. I have been fortunate to have received additional replies from him right into August 2017 as this book is about to be published. That same year as a result of an internet search for "time is motion" I discovered The Scientific Worldview web site, to which I began posting comments and several blogs, including one co-written with Glenn Borchardt, Ph.D.[1]

Because of my passion for learning about credible alternative scientific models I am a member of the Board of Directors of the John Chappell Natural Philosophy Society. Its website provides the following statement: "CNPS provides an open forum for the study, debate, and presentation of serious scientific ideas, theories, philosophies, and experiments that are not commonly accepted in mainstream science. The CNPS uses the term "Natural Philosophy" in its broader sense which includes physics, cosmology, mathematics, and the philosophy of science. Our goal is to return to the basics where things went wrong and start anew."[2]

Despite my enthusiasm for discovery, theories use invalid abstractions as occurs in orthodox physics and cosmogony, I am willing to challenge its proponents regardless of how esteemed they may appear. If a theory has contradictions, then it cannot be valid.

I review the Big Bang theory's multiple major problems, such as those of horizon, flatness, and monopole, conservation violation, evidence of no expansion, non-Big Bang theory explanations for the

Preface

CMB, inexplicable mature galaxies in the early universe, 1.65 trillion-year-old structures, contradictions from HUDF and GALEX data. In attempting to deal with its many problems, supporters shamelessly and frequently change the facts to fit the theory.

I discuss illusions in physics, such as matterless motion, wave-particle duality, superposition, probability waves, cosmological expansion; inflation, Schrödinger's cat, and the need of observers for the universe to exist. I argue quantum mechanics and relativity theories cannot be unified, because of their deep flaws.

I discuss invalidating problems in relativity, such as the non-scientific objectification of motion, violations of scientific assumptions, and profound math errors.

As alternatives to orthodox theories, I describe Glenn Borchardt and Stephen Puetz's Neomechanical Gravitational Theory, Borchardt's Infinite Universe Theory and Universal Cycle Theory; Stephen Bryant's Modern Mechanics theory; and Duncan Shaw's entanglement explanation and his aether gravity model. In a chapter on David Bohm's physics approach, I discuss causality and chance, pilot waves, the undivided Universe, Bohm's quantum potential (QP,) Bohm and Hiley's QP entanglement explanation, and Hiley and Callaghan's "Delayed-Choice Experiments and the Bohm Approach to quantum mechanics."

Other topics include a rational explanation for the double-slit experiment, Maxwell's aether, energy, dark matter candidates, and the cosmological principle, alternatives to galactic redshift interpretation, cosmological expansion and inflation myths, and emptiness myth. I posit that consciousness is an abstraction for brain processes such as thinking and feeling, and I account for the "hard problem" of subjective experience. I contend that microcosms in motion generate perceptions, which the brain interprets uniquely based on the mind's neuronal patterns.

I explain differences between awareness and consciousness, the cause of dualistic confusion, and describe the Global Workplace Theory. I list many questions about consciousness and the brain and provide logical answers. Notfinity Process examines the illusion of the observer, the concept of self-awareness, Antonio Damasio's model, and negative implications for "free will" from brain studies. Other areas concerning the consciousness abstraction include the evolutionary benefit of advanced mental processes, summary of neuroscience discoveries and higher–order thought theory.

Notfinity Process

My intention in writing Notfinity Process is to promote models in theoretical physics and cosmology that better represent aspects of reality, which exists independent of human presence. Although in challenging orthodox theories may seem to be an enormous task since they are so entrenched, I am inspired by the great physicist David Bohm. Using the analogy of the transformation of the atom ultimately into a power and chain reaction, he contended that an individual can transform humanity by applying their intelligence. Bohm thought that individuals had collectively attained the "principle of the consciousness of mankind." He held that an intense heightening of individuals who have shaken off the "pollution of the ages," which refers to incorrect worldviews contributing to ignorance, can start to generate the power needed to ignite the whole consciousness of humankind.

Delta, British Columbia, Canada
August 23, 2017
George S. Coyne

References

---

[1] Coyne and Borchardt, *Matter and Motion are Abstractions*.
[2] John Chappell Natural Philosophy Society.

Part One   Assumptions

Chapter I

Important Concepts

Focus and Theme

Reality involves a notfinity of process and relationship structures that persist while transient objects vary. That is one reason for the full title of this book Notfinity Process: Microcosms-in-Motion. Whereas quantum mechanics theorists contend that consciousness brings about wave function collapse to make our perceived reality manifest, I maintain that it is not that simple. Physicists have reached their conclusions by interpreting results according to their assumptions. Other assumptions, specifically the ten described in this chapter, can lead to alternative explanations, which do not have impossibilities inherent in them. To bring about advances in theory it is essential for physicists to get out of their "box." This can only be done by seeing that whenever contradictions or so-called "paradoxes" exist, the theory is invalid and one needs to look for different ones.

Meaning of Notfinity Process

Instead of the word "infinity," which we cannot truly capture in a concept, because it is beyond the confines of thought, I use the term **"notfinity"** as an abstraction for negating the possibility of finity. I

Important Concepts

intend the term to contain a reminder of the dangers in believing that any thought can represent more than what is finite. Because all concepts require thought, it is only possible to conceive of finite things. Thus, any concept of the Universe is by definition limited and finite. This is why I prefer to think in terms of what the Universe is not. It is not finite. When this term "infinity" originated in the 15th century, it meant "not finite," but after centuries of usage, it has assumed the additional meaning of something that one can hold in the mind as a representation of reality, rather than a negation of the finite. Notfinity refers to the absence of any possibility of finity, rather than an abstraction of infinity. "Universe" means limitless matter in motion. This univironmental, deterministic worldview involves causality, including behavior. Process refers to what all microcosms, which are all objects, are constantly doing.

Instead of the word "infinity," which we cannot truly capture in a concept, because all concepts are finite, I use the term "**notfinity**" as an abstraction for negating the possibility of finity. I intend the term to contain a reminder of the dangers in believing that any thought can represent more than what is limited. Because all concepts require thought, it is only possible to conceive of that which is finite. Thus, any concept of the Universe is by definition limited and finite. This is why I prefer to think in terms of what the Universe is not. It is not finite.

Bruce Nappi comments on this subject: "The human brain is an experience matching computer. When we try to imagine anything, the brain goes into its memories looking for a similar **streaming experience**. If it finds one, it plays it back. This gives us, not only the match, but a large amount of experience related to it. People do not have "rational" thoughts that represent more than what they have personally experienced! I added the term "rational" here because insanity and dreams can introduce unlimited hallucinations. Because of the circular feedback of the higher brain functions, the hallucinations and dreams are also recorded by the brain as experiences. ... So what does all this say about IN-finity? The problem people can't envision it in any form - i.e. the universe, notes going higher, colors beyond red and violet, pain beyond a hammered finger or burnt arm etc. is that they have no direct experience memorized that relates to the new realm. People blind from birth can never have the sensation of color because they have NO recorded experiences of color for future memory searches to find. Furthermore, human senses are totally incapable of recording perceptions that capture the "realm" of any infinity."[1]

When this term "infinity" originated in the 15th century, it meant "not finite," but after centuries of usage, it has assumed the additional meaning of something that one can hold in the mind as a representation of reality, rather than a negation of the finite. Notfinity refers to the absence of any possibility of finity, rather than an abstraction of infinity. "Universe" means limitless matter in motion. This univironmental, deterministic worldview involves causality, including behavior. Process refers to what all microcosms, which are all objects, are constantly doing.

There is no possibility of the term "finity" being in any way descriptive of the "Universe." This noun-verb refers to unlimited microcosms of limitless sizes, and their endless motions and interactions with their macrocosm and other microcosms. The term "notfinity process: microcosms-in-motion" expresses what I mean by this matter-motion word "universe."

In using the word "Universe" my abstraction goes beyond the idea of "totality" and "wholeness," and orthodox conceptions. Finite universe models do not seem to have any connection to reality. My concept of reality contains the understanding that there is no final, largest structure or complete collection of microcosms, nor is there any volume boundary. "Universe" cannot refer to an existing object or contained occurring process without being part of a macrocosmic structure, because objects and processes always require an environment. To think that an ultimate macrocosm exists by itself is ridiculous, because existence is only possible in an environment.

With no end to microcosms-in-motion, it is not possible to refer to the Universe's boundaries and thus it is more accurate to think of the Universe more in terms of limitless processes than an object. There is no ultimate macrocosmic universe per se. The "Universe" abstraction can only be instantiated by the existence/occurrence of individual microcosms-in-motion. Just as the word "food" is an abstraction for specified edible items, such as an apple, the same applies to the word "Universe." In both cases only the particular microcosm can be said to exist.

I have introduced some other new words in this book, such as **"nadal"** for the measurement commonly called time. This is a noun for the measurement of amounts of motion as indicated by clocks, rather than the actual movement, designated by "time." It represents the Earth's rotational motion. Rather than referring to the moving that is occurring, it specifies the movement as indicated by an object's positional change

Important Concepts

between each measurement. By measuring the motion of one microcosm's change in position relative to another microcosm, it is possible to ascertain a specific time, which since the first time-keeping devices has used the Earth's rotation as the comparison standard. "Nadalting" is a word that I created for the act of measuring time. Additionally, you will find some words defined differently than is the usual case. These modified definitions, such as for "time" used as a verb, are better able to represent the model of reality that I support.

One of the theories discussed in this book proposes that there is no limit to how large cosmic structures become. Although this does not contradict notfinity, the latter does not require holding ideas of infinity. It is simply an acceptance that there is no possibility that the Universe can stop existing, either beyond a specified smallness or largeness, or after any duration. Notfinity is negated finity. Using this abstraction may help us to realize that it is easier to negate finity than to imagine infinity, which inevitably involves finite conceptions because thought is limited.

Microcosms are all things and portions of the Cosmos; replacing the concepts of bodies, objects and systems. Process refers to what occurs with microcosms in continual motion.

My scientific worldview includes many aspects of David Bohm's holistic approach, classical mechanics when appropriate and systems philosophy. I have the view that thought patterns determine our experience of every situation, which in turn reinforces our pattern of thinking. This also applies to scientists with very few exceptions. For example, by dividing biology from physics in our thinking results in seeing the world in this framework, which makes one believe that these divisions are a given. All division only exists in our thinking. Reality is indivisible and holistic. Those who understand this will be a catalyst to bring about a new consciousness for humanity. The deeper one experiences the truth of this, the more that person will impact the entire world.

This worldview is based on the idea of univironmental determinism (UD). It states that every microcosm's evolution depends completely on the motions of matter within and without, and proceeds toward univironmental equilibrium. Because UD derives from causality, it assumes that there are an infinite number of causes for all effects. Determinism refers to the philosophical perspective that causality is universal; all events have physical causes that are determined by previous events, including all behavior, thoughts and choices. It has no similarities to predestination or fatalism, which denies the enormous power that we have to influence our macrocosm. Microcosms always affect the environment that contains them.

I describe the features of abstractions and reveal the differences between valid and invalid ones. Because a strong theoretical foundation depends on the validity of the underlying abstractions, having this information is crucial to assess the soundness of theories in physics and cosmology.

The invalidation of a cosmogony event, such as proposed by the Big Bang Theory, implies microcosms-in-motion have no limit. This can be proven by the fact that it is logically impossible for "nothing" to "exist." If what cosmologists refer to as "nothing" or "void" existed, then by definition it is something, because only things have existence.

The next chapter contains a description of the features of abstractions, revealing the differences between valid and invalid ones. Because a strong theoretical foundation depends on the validity of the underlying abstractions, having this information is crucial in assessing the soundness of theories in physics and cosmology.

I expose many problems and absurdities regarding the Big Bang Theory (BBT), as well as those in quantum mechanics. I describe problems in relativity theory, such as the non-scientific objectification of motion, violations of scientific assumptions, and profound mathematical errors discovered by Steven Bryant, sufficient to invalidate relativity. These theories constitute the deplorable state of orthodox theoretical physics and cosmology today.

As an alternative to these, I describe theories that do not have contradictions. These include Bryant's *Modern Mechanics*, which is a proposed replacement for relativity theory, classical mechanics and quantum mechanics. Also discussed extensively is Dr. Glenn Borchardt's *Infinite Universe Theory* (IUT) [2] contained in his book *The Scientific Worldview: Beyond Newton and Einstein* [3] and in the one he co-wrote with Stephen Puetz titled *Universal Cycle Theory: Neomechanics of the Hierarchically Infinite Universe*.[4] There is a thorough explanation of how time is motion rather than a flowing thing. I suggest the noun nadal to refer to the measurement of this motion. Additionally, there is a review of some of Duncan Shaw's profound papers. One is on Maxwell's aether and the other is an entanglement explanation that does not use quantum mechanics theory. Unlike orthodox physics theories, these consistent and intuitive models do not require one to abandon common sense and engage in Orwellian doublethink to accept them.

In addition, there is a chapter on the connection between brain, mind, self and consciousness. Recognizing the term "consciousness" as an abstraction for all the cognitive processes excludes the idea that this exists as a thing per se. I examine the "hard problem" of consciousness. This subjective experience results from matter in motion involved in generating perceptions, which the brain interprets uniquely based on the neuronal patterns that comprise the mind.

Some of the other physics and cosmology topics discussed include space is not a void, matter is an abstraction, energy is a matter-motion calculation; light is a wave in aether, Dr. Arp's galactic redshift interpretation—which does not imply cosmological expansion—and the illusion of a will.

Before astrophysicists and cosmologists will consider new theories for how the universe functions, they must be convinced that the established models consisting of quantum mechanics (QMT), General Relativity (GRT) and Special Relativity (SRT), are invalid. Despite the work of many scientists since the mid-1920s to reconcile these into a unified theory, there has been no success. This is because it is impossible to create a workable, coherent theory from deeply flawed ones that are not good models of reality.

If you are prepared to consider whether the contradictory and confusing theories of quantum mechanics, the Big Bang and relativity are invalid and need replacing, then you may find this book useful. It will assist you to discern between valid abstractions and paradigms in physics and cosmology from those that are not. Herein you will discover logical and plausible theories based on well-founded abstractions of the Cosmos and our inseparable connection to it. To succeed in having greater understanding of how the Cosmos works, you need to be prepared to abandon those theories that are no longer tenable, regardless of how much you identify with them.

Models that include impossible concepts have provided me with a source of enormous humor. Unfortunately, these inevitably confusing, contradictory theories are profoundly restraining the development of scientific knowledge and definitely not getting humanity any closer to the long-sought Holy Grail of a grand unified theory (GUT) or theory of everything (TOE).

Invalid abstractions do not represent what they purport to, and have no connection to reality. The results of these in theoretical physics have been disastrous. Contradictions abound in them, which indicate that they cannot possibly be correct or scientific. These absurd theories contain numerous inconsistencies and mutually exclusive ideas that orthodox physicists accept based on the assumption that the abstraction was well grounded and the theorists did the mathematics properly and it worked out. Another undeniable factor in the perpetuation of these theories over several generations is the massive support from the scientific establishment, including extremely influential Nobel laureate physicists, combined with severe negative consequences for those who dissent.

By learning to identify invalid abstractions, one stops believing that they are worthwhile and thus, no longer uses or accepts them. Increasing skill in this area empowers one in evaluating theories. This makes it less likely that the paradoxical theories that derive from these will sway or influence one prepared in this way. From this comes the confidence to reject theories that lack valid abstractions, regardless of how celebrated or respected the leading proponents of the theory are. If their reputation and status as an authority in physics or cosmology results from their work using untenable theories, that have already been falsified, then what value is their knowledge?

Asking challenging questions about the absurdities contained in theories, for which there are no rational answers, elicits different responses from scientists depending on whether the person asking the question poses a perceived potential threat to its continued acceptance. If the individual making the argument does not have a Ph.D. in physics, then orthodox physicists dismiss their ideas. Contrarily, some physicists, such as Richard Feynman, have stated that nobody understands QMT. By this reasoning, no one has enough understanding of it to be qualified to challenge it. There is a perfectly logical reason that no one grasps it, which has nothing to do with intellectual or perceptive powers. It is because it is so contradictory, paradoxical and illogical that it is impossible for a rational, sane person to understand how it can possibly be representative of reality.

Scientists who have credentials from prestigious universities, and have produced significant research papers sometimes attempt to publish criticisms and or alternatives to orthodox theories. Among physicists and in academia there is formidable resistance to the publication of their papers. Not having publishing opportunities for a university professor is

Important Concepts

obviously a major obstacle in their career. Even though the falsification process is an indispensable part of the scientific method, it takes a brave individual to be willing to attempt to falsify theories considered sacrosanct.

One very notable courageous astronomer, Halton Arp, who this book is dedicated to, questioned accepted theories about an expanding universe. Arp earned a Ph.D. from Caltech, did research at Mount Wilson Observatory, and was a fellow of the Carnegie Institution of Washington and an astronomer at Palomar Observatory for 29 years. Awards included The Helen B. Warner Prize for Astronomy in 1960, the Newcomb Cleveland prize for his address on "The Stellar Content of Galaxies," to the AAAS and the American Astronomical Society, and the Alexander von Humboldt Senior Scientist Award in 1984.

Despite his great expertise on the morphology and structure of quasars, soon after submitting a paper for publication that proposed an alternative astrophysical process to explain the problem of redshift discrepancies of quasars, he was not given telescope time and was compelled to resign from his job at the Carnegie Institution.

There is a high quality YouTube video *"Expanding Universe "Ryed shift" Myth debunked by Halton Arp using Quasars,"* [5] in which Arp speaks about his research on quasars indicating a non-expanding universe. He mentions how the astronomy establishment refused to consider his studies; Nature magazine would not even read his paper on quasars. Because his progressive scientific thinking, which means a dedication to the scientific method even though it does not support the Big Bang theory, the scientific powers sought to destroy his career.

His idea was that redshift is intrinsic, caused by the ejection of quasars from the center of nearby galaxies, rather than an effect of expansion, an idea that Big Bang cosmogonists consider unassailable. If this were correct then there would be no basis for an inferred expanding universe theory or Big Bang cosmogony.

In using this particular example, my intention is not to argue for Dr. Arp's ideas, as valuable as they are, but to propose that the extreme defensive reaction to his redshift paper by those with power demonstrates how original thinking is severely discouraged in attempts to protect the consensus theory. Because progress in science occurs through the development of better theories, scientists who offer alternative theories should be encouraged, rather than penalized.

The great astronomer Fred Hoyle, a good friend of Dr. Arp's, was was in a position to have good knowledge of how the professional order worked. His following statement is still applicable today: "The establishment defends itself by complicating everything to the point of incomprehensibility."[6]

QMT and relativity physicists, who are the most blatantly dismissive and even antagonistic to challenges to their theories, are behaving in an extremely unscientific manner. This sort of response, which is widespread amongst physicists, is more traditionally associated with religion than science. Many of the ideas and beliefs in the standard model defy rationality, logic and common sense more than any major religion that I know of. To facilitate progress in physics, orthodox theoretical physicists must change their attitude and stop penalizing scientists for expressing alternative theories.

I adhere to the scientific worldview. Merriam-Webster defines *weltanschauung*, which is the German word for worldview as "a comprehensive conception or apprehension of the world especially from a specific standpoint."[7] Hunter Mead defines it *as* "an all-inclusive worldview or outlook. A somewhat poetic term to indicate either an articulated system of philosophy or a more-or-less unconscious attitude toward life and the world."[8] All definitions generally include some reference to the individual or society's view having a quality of being "comprehensive," "all-inclusive," or "fundamental" in a wide range of areas.

Basic to the scientific worldview is the acceptance of the reality of determinism, meaning only matter can cause an effect. Because scientists define the universe as everything that exists, no "thing" can exist that is not some form of matter, including non-baryonic matter. Therefore, an imagined immaterial mind is not real, which excludes the possibility of any physical effects coming from this illusion.

In the 17th century Baruch Spinoza, one of the greatest known philosophers, argued the case for determinism in these statements:

> "All men are born ignorant of the causes of things, that all have the desire to seek for what is useful to them, and that they are conscious of such desire. Herefrom it follows, first, that men think themselves free as inasmuch as they are conscious of their volitions and desires, and never even dream, in their ignorance, of the causes, which have disposed them so to wish and desire."[9]

Important Concepts

"In the Mind there is no absolute, or free, will, but the Mind is determined to will this or that by a cause which is also determined by another, and this again by another, and so to infinity."[10]

As a determinist, I take the view that the interaction of the environment and system determines all events. I have held this perspective since 1974. This univironment, as Glenn Borchardt refers to it provides a model for understanding behavior: "the motion of one portion of the universe with respect to other portions."[11] Because I accept the need for a clear scientific worldview and approach to learning, I embrace the logically coherent ideas that Glen Borchardt formulated and delimited in *The Ten Assumptions of Science*[12]. This univironment, as Glenn Borchardt refers to it, provides a model for understanding behavior: "the motion of one portion of the As a determinist universe with respect to other portions."[17] Because I accept the need for a scientific worldview and approach to learning, I embrace the logically coherent ideas of The Ten Assumptions of Science[18] that Borchardt formulated and delimited. If you are interested in a more thorough explanation of them, you will find it in his book *The Scientific Worldview. Beyond Newton and Einstein.*

In my estimation, freedom does not lie in thinking that one's thoughts, including will, have no cause. Liberation from believing false ideas about the Cosmos and ourselves brings empowerment.

The scientific worldview refers to a combination of classical mechanics, which is about the inner workings of material things, and systems philosophy, which is about how things function outwardly. Borchardt explains a result of this: "Univironmental Determinism, simply states that the evolution of a microcosm is dependent on the motions of matter within and without. This evolution, this motion of the microcosm, is in all cases in only one direction, toward univironmental equilibrium. Being derived from causality, the assumption that there are an infinite number of causes for all effects, there can be no greater generalizations to be made concerning the mechanism of evolution."[13]

This means that an infinite amount and variety of matter in motion determines what transpires to any object, including life forms. Realizing one's part in this is very empowering.

Although we are not completely able to control what happens to us, our actions affect our life and the macrocosm in which we exist. The more that I live according to this worldview, the greater ability I have to be a determining influence. Borchardt expresses a similar idea very succinctly and clearly, in the conclusion of *The Scientific Worldview*: "Like the Industrial-Social Revolution, the scientific worldview discards traditional speculations, while opening new vistas to previously unattainable horizons."[14]

The Ten Assumptions of Science[15] Comparisons

No.	Determinism	Indeterminism
1.	Materialism	Immaterialism
2.	Causality	Acausality
3.	Uncertainty	Certainty
4.	Inseparabilty	Separability
5.	Conservation	Creation
6.	Complementarity	Noncomplementarity
7.	Irreversibility	Reversibility
8.	Infinity	Finity
9.	Relativism	Absolutism
10.	Interconnection	Disconnection

Paul-Henri Thiry, Baron d'Holbach demonstrated awareness of the assumption of inseparability and interconnection in stating:

> "Men always fool themselves when they give up experience for systems born of the imagination. Man is the work of nature, he exists in nature, he is subject to its laws, he can not break free, he can not leave even in thought; it is in vain that his spirit wants to soar beyond the bounds of the visible world, he is always forced to return."[16]

Philosopher Robin Collingwood[17] maintained that there are two main characteristics for fundamental assumptions: they always have opposites and it is impossible to prove them completely. When there are additional ones, then the two assumptions have to be consupponible, which means that you must be able to assume the other from either one without a contradiction.

The terms used in the ten assumptions of science are defined and described in the following section.

Important Concepts

1. Materialism

"The external world exists after the observer does not."[18] It is the assumption that there is a reality outside of our minds, which is not dependent on our observation for its existence. Because materialists are realists, they accept that matter is all that exists, and material interactions can completely account for all mental or psychological phenomena, including all thoughts and other aspects of consciousness and mind. In contrast, immaterialists as idealists assume that if reality has any existence outside of our mind, it is only as sensory impressions, and we derive reality within the mind. Bishop George Berkeley, an early advocate, in commenting on "all sensible objects," stated: "nor is it possible they should have any existence out of the minds which perceive them."[19] What he is clearly saying is that without observers, the Cosmos does not exist. This is preposterous and absurd in the extreme.

2. Causality

This is the assumption that there must be an infinite number of material causes for every effect because the Cosmos has micro and macro infinity,[20] which is the "eighth assumption of science." Merriam-Webster defines cause as "a reason for an action or condition."[21] Causality is universal in that all things are moving relative to all other objects and motions of all objects affect every object.[22] David Bohm brought enormous understanding of how determinism is workable. As Borchardt states: "Bohm showed that the quantum mechanical laws to which the Heisenberg Principle applies could not be assumed to be inviolate...Although we temporarily may be unable to find a material cause for a particular phenomenon, the cause nevertheless exists."[23]

3. Uncertainty

"It is impossible to **know** everything about anything but it is possible to have increased knowledge about anything."[24] Borchardt explains it as: "Uncertainty states that no matter how good our measurement of real objects, an improvement in precision is always possible, although perfect precision is not. There is always another cause that will explain part of the variation, but because the number of causes is infinite, some variation, some uncertainty will remain."[25]

4. Inseparability

As motion cannot occur without matter, so also matter cannot exist without being in motion.[26] Borchardt warns about four illogical mistakes that violate this assumption. "1. That matter could occur without motion. 2. That motion could occur without matter. 3. That matter is motion. 4. That motion is matter."[27] The category of matter includes the terms "thing,", "structure" and "mass"; whereas the abstraction of motion includes "event," "function," "velocity" and "time."[28]

5. Conservation

In stating this assumption, Borchardt avoids the misleading and confusing phrase of "conservation of energy." `It is only possible to conserve "things," and as energy is not a thing, the word conservation does not apply to this word. This is an important conceptual area to have clarity on. About the conservation assumption, he states: "Matter and the motion of matter neither can be created nor destroyed."[29] All occurrences are specific types of matter in motion transforming into other types of matter in motion.

Borchardt's use of the word "energy" to refer only to a matter-motion calculation differs considerably from the standard physics concept which results in differences in understanding conservation. Bruce Nappi explains the orthodox model this way "Energy" in any known form can convert to another form. Energy is "conserved" in these conversions. Matter, in any known chemical form (i.e. molecules, atoms) can convert to another form. Matter is "conserved" in these conversions. Matter elements in atomic form (protons, neutron, electrons) can convert into other forms (other atoms). Matter elements are "conserved" in these conversions. The MASS of atoms in nuclear reactions can change into energy. Energy in nuclear structures can change into mass. The sum of the mass plus energy in these reactions is "conserved."[30] It is important to keep this distinction in mind when reading this book.

6. Complementarity

All things are subject to divergence and convergence from other things[31] Borchardt refers to the paradox of the Second law of Thermodynamics (SLT). It states that the entropy of apparent disorder of an ideally isolated system can only increase. However, there is also the

Important Concepts

reality of increasing order in specific systems throughout the Cosmos. Philosopher–physicist L.L. Whyte referred to this in stating: "over vast regions of space and immense periods of time the tendency toward disorder has not been powerful enough to arrest the formation of this great inorganic hierarchy and the myriad organic ones."[32]

Bruce Nappi offers argues against including this as an assumption of science: "The hidden trap here is the lost definition of Entropy: the second law of thermodynamics. This law was developed to explain energy transformation in steam engines, and the behavior of gasses. Applying it to other sciences without clear clarification of terms has led to a lot of nonsense. It's like running tests on thousands of substance to observe their behavior going from solid to liquid. 99.99% of the time, the solid is more dense and sinks. But, that doesn't explain water, which does the opposite. So, proclaiming a Density Freezing Law and applying everywhere would fail."

7. Irreversibility

All processes are irreversible.[33] This arises from the deterministic interpretation of the second law of thermodynamics, including its complement, which states that the interactions of every object are unique. Because time is motion of things, all events or reactions are descriptions of motions and therefore cannot occur separately from time.[34] As the material objects in the macrocosm in which events or reactions occur are constantly moving, all reactions and events are irreversible. Reversibility would require completely isolating a system from the universe, which is impossible.

Nappi states: "there are challenges all over the place to entropy, when it is applied outside gas dynamics…some processes are reversible. For example, if water vapor in a bellows is subject to compression and expansion, it will transition from gas to liquid and back again—omplete reversibility. If a pendulum in an evacuated container swings on a knife edge bearing, it will swing with very small loss of motion, only due to bearing friction. However, the process of energy change between potential and kinetic is completely reversible on every swing." It is important to keep in mind that in the gas to liquid example provided by Nappi, when the liquid transitions back to a gas, the specific individual molecules will not be occupying their same original location.

8. Infinity

Borchardt states: "The universe is infinite, both in the microscopic and the macroscopic directions."[35] The reason for this assumption is that infinity is a requirement of the causality and uncertainty assumptions and is the only form compatible with these.[36] An explicit infinity assumption exists in Bohm's explanation of universal causality, because it is impossible to have absolute certainty of the cause of an effect.[37] Those who believe that the universe had a beginning are rejecting physical causality, because a beginning implies that there was no matter in existence to cause the universe to appear.

9. Relativism

"All things have characteristics that make them similar to all other things as well as characteristics that make them dissimilar to all other things."[38] Making comparisons between things is fundamental to the method used in thinking and forms a basis for making scientific statements and all other statements.[39] In this assumption, the concept of absolute equalities or inequalities is invalid. This is because of the constant motion of all things, resulting in continual changes to everything and therefore more differences compared with other things.[40] Crucial to comprehending relativism is having an understanding of the use of classifications, which means there needs to be agreement on comparing the identical measurable features.[41]

Stellar halos are diffuse and faint components of galaxies that offer cosmologists the opportunity to learn about galaxies' assembling histories. Max Planck Institute for Astrophysics (MPA) research scientists discovered that galaxies with similar luminosity, mass, and morphology have great diversity in halo properties.[42] This demonstrates the ninth assumption of science.

10. Interconnection

"All things are interconnected, that is, between any two objects exist other objects that transmit matter and motion."[43] Although within every microcosm there is continuity, this assumption includes the idea that there is no absolute continuity within anything. Additionally, there is no absolute discontinuity between an object and its environment, because every microcosm and its macrocosm have things in common. Borchardt points out that the assumption of interconnection excludes the possibility of completely solid matter or empty discontinuous space.[44] This assumption includes seeing a connection between objects of any size, such as between stars, galaxies, clusters of galaxies, galactic superclusters etc. There is always some form of matter between these microcosms providing a medium for interconnection. If aether theories are correct, then this medium is mostly aether.

In the *Neomechanical Gravitational Theory* (NGT), found in Chapter 12, assumptions 4 and 5 are important, but assumptions 6, 8, 9, and 10 are of crucial significance.

References
---

[1] Nappi, B. in e-mail to me.
[2] Borchardt, *Infinite Universe Theory*.
[3] Borchardt, The Scientific Worldview.
[4] Puetz and Borchardt, *Universal Cycle Theory*.
[5] Youtube, Expanding Universe "red shift" myth debunked by Halton Harp using Quasars.
[6] Hoyle Quotes by Fred Hoyle.
[7] Merriam-Webster, *Definition of Weltanschauung*.
[8] Ken Funk quoting Mead *in What is a Worldview*.
[9] Gary Zabel. *Ethics Part One Concerning God, Appendix*.
[10] Stanford Encyclopedia of Philosophy, Quotes *Baruch Spinoza*.
[11] Borchardt *The Scientific Worldview*, xxvi.
[12] Borchardt, *The Ten Assumptions of Science*.
[13] Borchardt, *The Scientific Worldview*, 157.
[14] Ibid., 320.
[15] Borchardt, *The Ten Assumptions of Science,* 10.
[16] *Today in Science History* , Quotes by Paul Henri Thiry, Baron D' Holbach.
[17] Borchardt and Puetz cite Robin Collingwood. *Gravitation Theory.*
[18] Borchardt, 17.
[19] Berkeley, quoted in G. Borchardt, *The Ten Assumptions of Science*, 16.
[20] Ibid., 28.
[21] Merriam-Webster. *Cause*
[22] Ibid., 21.

[23] Ibid., 26, 27.
[24] Ibid., 31.
[25] Ibid., 35.
[26] Ibid., 47.
[27] Ibid., 60.
[28] Ibid., 57.
[29] Ibid., 63.
[30] Nappi, B. in e-mail to me regarding conservation in Notfinity Process.
[31] Ibid., 76.
[32] Borchardt quoting L.L. Whyte quoted in *The Scientific Worldview*, 73.
[33] Ibid., 79.
[34] Ibid., 87.
[35] Ibid., 88.
[36] Ibid., 91.
[37] Ibid., 27
[38] Ibid., 98.
[39] Ibid., 106.
[40] Ibid., 100.
[41] Ibid., 100.
[42] Max Planck Institute for Astrophysics, *The diversity of stellar halos in massive disk galaxies.*
[43] Ibid., 107.
[44] Ibid, 108.

Part Two **Abstractions**

Chapter 2

Characteristics of Abstractions

Understanding their Meaning

The problem of contradictions and paradoxes in theories goes back to the abstractions used in constructing the model. That is why I have taken many pages to discuss abstractions. To have increased skill in discerning between valid and invalid abstractions in theoretical physics, it is crucial to have a thorough understanding of their features. This will help one to see that appropriate ones always relate to the real world. This chapter provides important information on this topic including ways to recognize ones that do not represent reality. While reading about their features, it is useful to think about how they relate to those used in theoretical physics and cosmology theories.

Abstraction comes from "abstrere," which means to draw away from things. While an abstraction is a concept, it also links all similar concepts as a category, field or group while serving as a super category that contains concepts. It comes out of a process of deriving concepts and general rules from methods, such as using basic, self-evident assumptions or propositions known as first principles. These concepts link particular examples, words and other images, sounds, and symbols representing an underlying meaning or concept. Because abstractions look for similarities, they have a tendency towards generality. In this book, I am using the tem first principles as referring to those that arise from well-established science.

Jennifer Vonk, Ph.D. is a comparative/cognitive psychologist with research interests in animal cognition, and cognitive development. Her studies suggested that gorillas and orangutans are capable of category discriminations of an intermediate level. She concluded that the experimental results provided evidence of abstract representation that is conceptual as opposed to simply perceptual.[1]

Although other species, such as chimpanzees, dogs, and bears, appear to have an ability to recognize categories, only people use and understand highest-level abstractions. As Noam Chomsky, often described as "the father of modern linguistics," has maintained during his lengthy career, all human thinking involves some degree of abstraction, because this capacity is innate.

This capacity is crucial for present-day society to function. Without the growth in this ability, highly abstract language and advanced mathematics would never have developed, which itself contributes to additional development of aptitudes in abstract thinking.

This chapter mentions examples of abstraction that have relationships to theoretical physics, used in philosophy, mathematics and computer science. The mental objects that comprise our consciousness are generally abstractions, most of which are both useful and essential, but others are deleterious to our health and happiness and to that of other people who we are associated with directly or indirectly. These harmful abstractions can prevent understanding when we forget, or do not see, that all of them are merely representations for something.

Abstract versus concrete objects

Whatever has physical referents is termed a concrete object or concreata whereas abstract objects, known as "abstracta," lack any physical examples in a specific place or time. An object is considered to be abstract if it lacks any ability to affect anything causally. Because these abstractions can never be instantiated, as they lack tangible referents, they are only mental constructs. This particular idea is extremely relevant when we get to the section on problems with the standard model in theoretical physics and cosmology. For example, the idea of Minkowski "spacetime," which is used in relativity theory is entirely a visualized mental image. Knowing what qualifies as abstract things facilitates evaluating the theories associated with these disciplines.

Although all abstractions are categories rather than things, any particular object in the category is a real thing not an abstract thing. Nappi points out that "abstracts ultimately point to concretes, even if there may be multiple levels. For most abstractions, it is often necessary to proceed through multiple levels of abstraction before reaching a concrete referent that is not a product of thought.

Nappi comments: "At each level of abstraction, there must be a defintion of the categorization. The definition is a logical MAP! If all terms are held strictly to their definitions, then there is a logical connection between the category "word title" and the terms it points to. Each of those terms, in turn, will point to others further down. Eventually, there is a logical path between all the levels and a set of root concretes.... The reason abstracts are needed in the first place is that pointing to a concrete item leaves out the whole issue of function! When a group of trees is called a forest, it's not just a group of trees within a boundary. The grouping develops functional characteristics that a single tree doesn't have. Each abstraction must be part of a logical MAP that can include other accurate abstractions, and true relationships that ultimately point to accurately empirically observed tangible entities.[2]

Features of Abstractions

The information in this section on the properties of abstractions is very important to understand because all theories use simple and complex abstractions to create higher level ones. Having a solid foundation in understanding differences between valid and invalid abstractions is essential to evaluating the theories that depend upon them. This greater knowledge on the features of abstractions will assist you in realizing how abstractions are able to be powerful communication tools when applied properly. You will also see the dangers associated with invalid ones, and in improperly using concepts in the category referred to as types, such as confusing compound and discrete ones. I explain this further in the Application of Types subsection in this chapter. In *Disruptive*, Steven Bryant demonstrates how Einstein made this mistake, which resulted in his relativity theories being untenable and invalid. I discuss this in Chapter 6 in the section on Einstein's Errors. The features include instantiation, compression, referencing, material process, cognition, existence status, simplifying, and physicality.

Instantiation, "as in a creation of an instance or occurrence of a concrete thing"[3] is a representation of an abstraction by a concrete physical thing. For example, athletes can be instantiated by a particular basketball player. The principle asserts that instantiation is required at some point in time for any property to exist. Also known as the principle of [8]exemplification,` it affirms that it is not possible for a property of something to exist without the object. The color blue does not present by itself; it is always an aspect of something that exists. This may seem obvious, but theoretical physicists ignore this fact every day in their work. When evaluating the validity of abstractions in physics it is useful to keep in mind the necessity of an actual object in order for properties to exist. For example, as motion is always an aspect of what things are doing, the relativity concept of motion without an object in motion is absurd. If anyone were to question physicists on how motion could possibly be separate from the thing in motion, they would say that they have the mathematical equations for it, failing to see that math is pure abstraction, not a real-world 3-D experimental result they can replicate in a lab.

This abstraction of motion occurring on its own has so thoroughly permeated into their thinking from years of indoctrination and habitually using it in physics equations, that they do not realize that it is an abstraction without a concrete referent existing apart from the moving object. They believe it is something that is part of the universe as much as matter is. Although matter, as meaning everything that exists, is also an abstraction, it is possible to instantiate it with a concrete referent, whereas with motion this is never possible at any level. In several areas of this book, I discuss the profound confusion that physicists have about motion.

To be multiply instantiated means that there are constituents or specific examples of an abstraction in varied times and locales as in the abstraction of extinct species. Thus, even if an abstraction has no present concrete referents, the abstracted property is an existing abstraction provided it had previously been instantiated. Applying this principle to any given abstraction, an abstraction only refers to real things or occurrences when one of them has been or is presently instantiated. If, regardless of how many levels of abstraction moving towards the less abstract one considers, it is never possible to come to a concrete referent, then the abstraction cannot be considered valid and therefore should not be used in any theory, especially not in ones on how the universe works. For example, a specific dog is an instantiation of the abstraction of "dogs."

The preceding information on instantiation is crucial to discerning whether abstractions are valid categories that contain real things and genuine occurrences, or those that only contain abstract things. The latter never contain any "thing" that is not the product of thought. Keep this concept in mind when evaluating relativity theory (SRT and GRT) and quantum mechanics theory (QMT

In forming abstractions, **compression** uses resemblances of several things to create one abstract item. An example of this is mapping millions of cars to the "car" abstraction. Confusion in distinguishing what is a concrete thing from that which is abstract does not occur, because one focuses on an innate equivalence of abstract and constituent data. Beginning with sensory data reception, we can infer abstractions from the most concrete level to increasingly larger categories. Here is an example using a "human activity" abstraction starting with Aaron Rodgers holding a football. One can move to the abstraction that he is a player for the Packers, which is a referent for an NFL team. In turn, this is an example of professional sports organizations, which abstracts to athletics. Proceeding to greater abstraction, athletics is a referent of exercise, which represents human activity. This abstracts to a category of all animal activity etc.

Referencing means precision in an abstraction's referent, as many have ambiguity on this, depending on whoever uses it. Friendship, honesty, and kindness are examples that signify different things to people when applying the variables of age, cultural background, personality, beliefs and numerous other factors associated with the individual employing the abstraction.

Referencing problems are plentiful in orthodox theoretical physics. There is enormous ambiguity due to a lack of adequate definitions from physicists who accept SRT and GRT. This applies to abstractions such as matter, dark matter, energy, dark energy, time, spacetime, motion, momentum, force, gravitation, fields and many others. Physics Nobel Prize winner Richard Feynman spoke about the referencing problems associated with energy: "It is important to realize that in physics today, we have no knowledge what energy is."[4] He explained that physicists do not have a concept that energy exists in "blobs" of a defined amount.[5] Similar referencing issues are associated with many other terms commonly used in theoretical physics. When encountering these, it is useful to question the validity of the abstraction.

In going from abstract to concrete referents, there are often multiple levels of abstraction known as ordering. For example, we can start by considering thought at its most abstract level as an idea that one seeks to convey. We specify it as "communications", and use the abstraction of "the media", which is instantiated as the broadcast form represented by television programming. We then specify it as a documentary program on science and so on with each step becoming increasingly more specific.

Material process involves a moving away from the concrete to the abstract. We see this in the development of democracy concerning decision-making in public affairs. It involves transforming from a pure democracy abstraction involving direct voting by all citizens on all issues, to an indirect form that uses representatives. In theory, these elected officials represent their constituents' views, but in practice, their votes are far more reflective of big financial contributors who begin lobbying as soon as the election is over. Thus, the abstraction for democracy morphs into an abstraction of "money equals political power."

In theoretical physics, material process involves drawing away from an observation of a particular phenomenon, such as quasars discovered in the 1950s, and forming an abstraction of it. Once a concept of quasars as "compact areas" centered in a massive galaxy surrounding a central supermassive black hole is the accepted explanation, then inferences about them spring up. For example, that they are 9 billion to 10.5 billion light years from us based on interpreting their redshifts in terms of Hubble's law. Thus, by a process of selecting redshift data from all that is available coming from quasars, and using the assumption that this represents the recession speed of matter from us, scientists believe they can determine a quasars' distance.

However, if the hypothesized expanding space between the quasars and us is a misinterpretation of what transpires, it cannot be contributing to redshift. Then the Big Bang abstraction based on a supposed cosmic expansion using redshift as evidence of it would be invalid, as would the abstraction formed relating to quasars. This demonstrates how crucial the feature of material process in abstractions is.

Abstraction involves generating ideas and distancing them from the thing or process connected to the abstraction. This feature of creation of ideas involves **cognition**. In this, if the first assumption is wrong then any theory depending on it is meritless. For example, in the debunked 1800s phrenology theory, practitioners assumed that a variety of traits and aptitudes, such as intelligence and musical ability, were associated with specific areas of the brain. They conjectured that the cranial skull conforms to the various sizes of these brain areas. From this idea, phrenologists concluded that they could ascertain an individual's capacity for particular psychological characteristics through skull measurements.

Using this erroneous assumption, phrenologists proceeded to create abstractions including maps of 27 areas that supposedly represented the connection between specific parts of one's brain and types of personality and behavior. In the 1930s and 1940s, the Nazis used invalid phrenology as a bogus "scientific" basis for the eugenics movement to provide the appearance of a basis for prejudice. One can make a strong argument that it led to the promotion of genocide by using it in a way that was supposedly "scientific."

Posing questions on **existence status** helps in discovering whether we are ascribing realness to properties and ideas versus that of material objects in the universe. This aspect of abstraction is enormously relevant to theoretical physics where there is confusion about these distinctions to the point that the belief in real existence attaches to many concepts that are simply ideas. For example, relativity physicists in modeling the universe, include the Minkowski spacetime abstraction. However, there is no instantiation for this concept. In this theory, time is the fourth part of the spacetime continuum in which everything exists. In developing, a theory of the Universe there is no need to make time into a dimension in spacetime. As Steve Bryant pointed out to me, this error originated with Einstein, who "mistook a duration as a replacement for Newtonian time (or universal), which still applies."[6] Instead, we can refer to a process of microcosms (i.e. arbitrary portions of the universe, which includes all things[7]) moving and changing position relative to other microcosms. An abstraction's existence and occurrence status is apparent in real things and events.

Although appropriate abstractions have great value in theoretical physics, it is of crucial importance to keep in mind that improperly formulated abstractions are of things that do not actually exist. This occurs when theorists attempt to combine space, which is actually a form of aether matter, with time, to fabricate the purely imaginary spacetime.

Because spacetime is an abstraction without a referent to any real thing or occurrence, questions about its ontology are appropriate. Today's orthodox scientists no longer refer to a complete vacuum. Now they talk about the void having fields, zero-point energy and virtual particles, which immediately annihilate upon their appearance in the vacuum. Large masses supposedly warping immaterial spacetime is the standard explanation for gravity and the expansion of space between

Characteristics of Abstractions

galactic clusters. This does not explain why the universe supposedly grows larger at an ever-increasing rate. One may logically wonder how space, which relativity theory considers as non-material, can bend and expand.

Existence status is also applicable to abstractions in QMT, such as the idea of quantum superposition. Many sections of this book refer to this outrageous and implausible idea.

In Causality and Chance, David Bohm uses "abstract" to mean "taking out" commonalities to a variety of similar things, and he clarifies that all abstractions, involve **simplifying** something by conceptually removing it from its complete context.[8] In simplifying, some degree of obscurity in abstract communications such as speaking and writing—replaces what was previously explicit. Therefore, to know what the communicator seeks to convey, the listener or reader needs to have some amount of shared experience and familiarity with what the abstraction designates. People generally accept the loss of precision for the benefit of the ability to communicate necessary concepts.

Physicality is another common feature of abstractions. An example of this occurs in the purchase of paper gold futures for the right to buy a specific amount of gold at a set price and date. Because almost everyone closes the contract, rather than take delivery of the physical gold, a 293:1 ratio of paper to gold available for delivery has developed. With more contracts standing for delivery than available gold as closing dates approach, to avoid a failure to deliver, the COMEX offers cash incentives for settling in cash. This situation developed due to the distancing of an abstraction from the thing that it originally represented. The significance of the cash incentives is that the stated price for gold is not actually representing the true price that one can obtain for gold due to the disconnect between the huge 293:1 ratio at COMEX in sales to supplies.

Fully comprehending the physicality feature requires knowing that abstract things have no existence in the real world of concrete objects. They are simply a way to refer to one's sensory experience. Further discussion on this topic is in the next section on philosophers. Understanding physicality is crucial when evaluating abstractions used in QMT, SRT and GRT. For example, in QMT, the word 'wave' signifies abstract wavelike properties without any matter, yet physicists conceive of these "waves" as having an existence in the real world.

Having greater appreciation of the meaning, significance, and implication of abstraction characteristics, such as instantiation, compression, referencing, material process, cognition, existence status, simplifying, and physicality, increases one's skill in discerning the validity of concepts used in scientific theories.

Application of Types

Using the concept of "types" enables one to classify things to prevent improper use of abstractions, (e.g. reaching a sum from adding two different species) thereby protecting us from reaching wrong conclusions, making errors in math or having invalid concepts. We need to know the rules applicable to types to use them appropriately so that we can avoid creating invalid abstractions.[9]

The two categories of types are "discrete" and "compound." Discrete types are static values and remain constant regardless of any change in other values. For example, the time variable does not affect the distance between two goalposts. In the constancy of time, motion that has occurred does not affect the measurement of the amount of motion that is occurring.

To measure the movement of objects through space, such as Earth's rotation and orbit around the Sun, it is necessary to create an abstraction of the movement. That abstraction is a discrete type that confusingly uses the word "time" to refer to a measurement of motion as indicated by clocks and to a thing that supposedly exists as part of the spacetime structure.

The distinction is very important because physicists often confuse measuring time with the movement that is time. This error is tantamount to thinking that a measurement of a building's height is the building itself. Whereas architects work with lengths every day, they don't confuse these numbers with what they represent in the structures they are designing, unlike physicists working with measurements of time who mistakenly believe the measurement is time. As Bryant discusses in *Disruptive*, three principal discrete types exist.[10] Distance is the measured length of space or other things. Time, when used as a measurement, refers to constant intervals of motion. One can avoid confusion through realizing that although one of the common meanings of the word "time" is measurement, it is incorrect to think that this calibration is equivalent to the motion that occurs. It is thus necessary to distinguish between time as motion, which refers to a process involving the change in position of an object, and the measurement of that change in position. Therefore, a new term is required because the word "time" cannot appropriately designate both. I propose the noun "nadal" to represent this measurement of motion. Whereas other works use mass, velocity, or something else as a third type, Bryant uses cycles, which is a sequence of changing states that, upon completion, produces a final state identical to the original one[11] as a third type.

Using these three discrete types, one can construct the compound types: frequency, velocity and wavelength. Frequency is the total occurrences of a cyclical event per unit of measured "time". Velocity, which is a function of time, is the rate of change in an object's position relative to a frame of reference. Wavelength applies to standing, traveling and other spatial wave patterns. It refers to a length per cycle, which is the distance over which a wave's shape repeats, and the inverse of the spatial frequency. Looking at the distance between consecutive corresponding points of the same phase (i.e. zero crossings, crests or troughs.) establishes what the wavelength is. A compound type, such as kilometers per hour for speed, refers to distance traveled related to a specified percentage of one Earth rotation. The type of the numerator is what proceeds "per," and type of denominator follows it. Dividing any type by the number one inverts it, such as inversed time or inversed frequency, etc.

Serious errors result from not recognizing differences between discrete and compound types. These include using the compound wavelength in place of the discrete length. As Bryant discovered, Einstein made this kind of error while formulating Relativity Theory. When he is actually referring to length he confuses it with time (as in light-year, which assumes knowledge of light velocity), and for wavelength, he uses a number for length instead of length per cycle.[12] This is a major problem for Relativity Theory. In addition, as Bryant found in Einstein's spherical wave proof, which was the key one used as a proof for relativity; there was a type I error that is sufficient to invalidate the theory.[13] I explain this in Chapter 6 in the subsection "Einstein's Errors."

Abstractions used in Professions

For philosophers, the abstraction process involves creating a concept of a particular characteristic shared by many people by isolating it. In attempting to determine the concreteness of many concepts, such as honor, courage and compassion, predicate abstractions are used. Propositions about predicates are questions that one poses concerning properties of the concepts. These are useful to explore a concept's abstractness or realness.

This process facilitates a greater appreciation concerning the nature of properties referred to as "universals" in terms of whether they exist in the real world or just in thought, and if they do exist, what they are. It also helps to explore problems of whether sensory experience is the only or principal source of all knowledge.

Using propositions about predicates would be useful to examine an abstraction, such as Quantum Field Theory in curved spacetime. This theory contends that gravitational fields generate particles. These fields, which can have horizons or be time-dependent, are another way of referring to curved spacetime in GRT.

Philosophers debate whether universals represent things that exist in reality or are only thoughts within the mind. If they have concrete existence, then it should be possible to instantiate them in the real world as a separate thing from what they are associated with. If it is determined that this is impossible to do, that implies they are only structures of thought. Applying universals to theories in physics would provide an additional way to evaluate them.

Programmers use abstractions for conceptualization that comes with the development of programming frameworks and libraries, which can enable reusing models written in a human-readable source code. Translating this into machine language makes it accessible to all computers. Computing difficulties are resolved by sequestering concepts and categories from particular occurrences of execution. Based on an abstract concept of the problem's solution, rather than precise details of the operating system hardware, software and support applications, the resulting code is easily incorporated.

Mathematicians do abstractions by finding a concept's indispensable qualities and isolating these from anything they were once associated with in reality. For greater matching or increased applications to abstract depictions of comparable phenomena, they need to derive general principles. Abstraction enables mathematicians to apply methodology used in one field to related fields for proving results.

Mathematicians minimize ambiguity, the primary problem associated with abstraction, by applying highly specified rules. This facilitates constructing complex equations and generalized models, which can have extensive application in approximating reality and the way the universe functions. Although math has high precision and ability to work in particular theories, this does not imply that those theories are accurate or inaccurate representations of reality. The limiting factor for mathematicians in fully incorporating highly abstract concepts is that abstraction requires much insight gained from frequent use of math structures.

References

---

[1] Wonk, *Matching based on biological categories in Orangutans. (Pongo abelii) and a Gorilla (Gorilla gorilla gorilla)*.

[2] Nappi, B.in e-mail comment to me on abstract vs. concrete objects.

[3] Ibid.

[4] Feynman, volume I; lecture 4, *Conservation of Energy*; section 4-1, *What is energy*.

[5] Ibid.

[6] Bryant, in correspondence with myself.

[7] Borchardt, *The Scientific World*view, 121.

[8] Bohm, Causality and Chance, 2

[9] Bryant, Disruptive: *Rewriting the Rules of Physics, 123.*

[10] Bryant, Disruptive.

[11] Merriam-Webster, *Definition of Cycle*.

[12] Bryant, *Relativity Challenge.com*.

[13] Bryant, *Disruptive* 13 to 29.

Chapter 3

Appropriate Role of Math in Physics

Views from Scientists

In 1623, Galileo stated: "Philosophy (nature) is written in this grand book, which stands continually open before our eyes (I say the 'Universe'), but cannot be understood without first learning to comprehend the language and know the characters as it is written. It is written in mathematical language." Commenting on this same idea during a lecture in 1921 at the Prussian Academy of Sciences Einstein asked: "How can it be that mathematics, being a product of human thought which is independent of experience, is so admirably appropriate to the objects of reality?" Einstein answered this question with the statement: "As far as the laws of mathematics refer to reality, they are not certain; and as far as they are certain, they do not refer to reality."[1]

A huge error in Einstein's statement (in which he appears to refer to mathematics as a product of human creation) is in stating that thought is 'independent of experience," which would be impossible.

Also in *The Unreasonable Effectiveness of mathematics in the Natural Sciences,* physicist Eugene Wigner thought we must find out why physics was inherently mathematical. In connection to this question mathematician, physicist and astronomer Sir James Jeans, who believed that God's mathematical thought was the source of the universe had a comment on this. On page 134 of *The Mysterious Universe* (1930) he stated: "We have already considered with disfavour the possibility of the universe having been planned by a biologist or an engineer; from the intrinsic evidence of his creation, the Great Architect of the Universe

Appropriate Role of Math in Physics

now begins to appear as a pure mathematician."[2]

Richard Feynman believed that mathematics was the best way to represent and understand the natural world: "To those who do not know mathematics it is difficult to get across a real feeling as to the beauty the deepest beauty, of nature. ...If you want to learn about nature, to appreciate nature, it is necessary to understand the language that she speaks in."[3]

Steve Bryant sent me an interesting response to this comment. He states: "I disagree with Feynman. I don't think nature speaks in mathematics. But, mathematics is the paintbrush we use to capture our impressions of nature. Nature is no more math than nature is a painting."[4]

Isaac Newton's *The Mathematical Principles of Natural Philosophy* (1687) launched math's role in physics of quantifying things and then discovering relationships between those numbers to reveal interactions between forces, masses, and his three laws of motion. This is still how physics functions in its pursuit of a model for the interactions between elementary particles, though now sophisticated abstractions, such as matrix algebra, group theory, symmetry, and probability waves, aid this endeavor. Absolutely proving models is not possible, but falsifying them is achievable. This occurs when observations contradict a theory.

Physicists discovered that equations could model the universe's laws in its patterns and shapes, which include electromagnetism, gravitational effects, radiation, chemical interactions, and other particle motion. Physicists have moved from just seeing connections between aspects of nature and their formulae to thinking that the math was representing the whole universe. This belief in the power of mathematics has led to a century long search for the theory of everything *(TOE)* or a unified field theory. *(UFT)*. For many years, physicists have fallen into the view that math is more than just a way to represent reality, thinking that it is the very "essence of reality." Physicist Max Tegmark's book,[5] *Our Mathematical Universe: My Quest for the Ultimate Nature of Reality* contends that all reality is merely a mathematical structure.

Mathematicians may come to think that their abstractions have some existence apart from thought. Forgetting that their discipline is for finding and proving relationships, they may mistakenly claim, for instance, that mathematics has proven a certain QMT result.

Physics uses mathematics in an amazingly coherent and logical internal system to describe math relationships involving concrete concepts. External relations set by the natural world in conjunction with observations and experiments determine selection of mathematical relations. QMT's attempt to be the connection between abstraction as math and the real natural world is a colossal failure.[6] Consider the significance of this quote by J.D. Jackson: "As indicated at the end of the introduction, such idealizations as point charges or electric fields at a point must be viewed as mathematical constructs that permit a description of the phenomena at the macroscopic level, but that may fail to have meaning microscopically."[7]

Concurrence between observations and predictions does not offer definite proof that an interpretation is right as a different one can produce the same or a similar result. However, this does not mean its proponents will stop supporting it, as is demonstrated in the Big Bang Theory, which has many observations that contradict it.

The internal relations of physics use only mathematics, but physics' external relationships involving coherence with the natural world–as expressed in the Copenhagen interpretation and relativity theory–violate logic and rationality. This relation between mathematical formalism and nature has never been resolved. Great bewilderment has developed due to the fundamental distinction between the role of causality in physics and that of logic in mathematics.[8]

Language of Math

Mathematics identifies and distills many indispensable features of languages. Therefore, any form in mathematics is just an adaptation from what is already included in language. Math is the only language for physicists to communicate with and produce accurate measurable predictions.[9] Progress in physics occurs when the content of concepts and math expressions are well-matched, enabling effortless communication while pointing a theory in a specific direction.[10] The codified form in mathematics enables it to exceed languages' capacity. This facilitates performing calculations, making conclusions and showing proofs, while requiring visual-sensory motor thinking not related to verbal language, but using a direct, internal visualization. Math's relationship to physics has surpassed its earlier supportive role to one of dominance. Mathematics is not only able to connect specific data with abstractions, but also create greater abstractions to model how these abstract relationships relate to each other, and at even higher levels, it uses category theory to represent relationships between different fields.[11]

Limitations of Math in Theories

The utility of abstract mathematics in making accurate predictions is undeniable, but the amazing predictive power that math offers is a double-edged sword, because established math is elevated to a position that is not justified. The most perceptive mathematicians and physicists understand that although it is possible to make mathematical mental constructs to represent reality, this does not mean that an idealistic view has any validity.

Albert Einstein commented on this on January 27, 1921 in an address titled *Geometry and Experience*: "One reason why mathematics enjoys special esteem, above all other sciences, is that its laws are absolutely certain and indisputable, while those of all other sciences are to some extent debatable and in constant danger of being overthrown by newly discovered facts. In spite of this, the investigator in another department of science would not need to envy the mathematician if the laws of mathematics referred to objects of our mere imagination, and not to objects of reality... But there is another reason for the high repute of mathematics, in that it is mathematics which affords the exact natural sciences a certain measure of security, to which without mathematics they could not attain."[12]

Some of math's deficiencies, when applied to theoretical physics, are its ambiguity and limited strategies for addressing these ambiguities. Not having present math structures to accommodate new theory hinders its development. This results from overemphasizing math's role in physics. Misinterpreting math or using it in a way that is incapable of representing anything beyond an abstraction can be problematic. Yet the scientists look at the equations and results as being equivalent to reality.

Although it should be obvious, many physicists do not seem to realize that just because there is a match between what they observe and what the math predicts is not proof that a particular interpretation is right. This is because other interpretations may give a similar or the identical result.

Glenn Borchardt made an excellent point regarding math in this statement: "Mathematical derivations may be internally consistent, but they do not necessarily have anything to do with the real world."[13]

This is the case when physicists accept a theory as being valid regardless of how non-intuitive its explanations may be, as long as the equations work out.

One of the best examples of this is the theory of mathematician and geophysicist Sydney Chapman. Due to being mathematically elegant, it was accepted over the competing theory, which was actually the correct one based on observation. This is discussed in Chapter 4 in the section on Hannes Alfvén. In addition to the concept of spacetime many other examples of mathematics not referring to the real world are found in relativity theory.[14]

This overvaluation of mathematics has led physicists to accept that if a theory's equations produce the correct prediction then even the most obviously absurd one must be representing reality accurately.[15] Realizing the ridiculousness of their model, they then conclude that reality must be absurd. That improper reasoning is something every physicist should keep in mind. If physics causes the particular math form to be difficult to follow, this stifles creativity and progress.[16]

Theory development is being hindered by physicists not considering ways in which their thinking and expression is circumscribed by the medium of the math forms being employed, thereby interfering with further progress in developing theory.[17] Problems with an overemphasis on math arise by requiring all theoretical physics abstractions to fit math's form. Rather than continuing the trend of physics being subservient to mathematics, I maintain that this is not the way to proceed. Math needs to be a tool for physics, not its master. Creating theory and doing science involves far more than math, which is an essential tool, but it is foolish to allow it to be more than that.

Borchardt's Infinite Universe Theory is free of contradictions and is not paradoxical, but as he has stated, it is not fully compliant with mathematics. This is unnecessarily limiting this logical, rational intuitive theory from gaining the amount of consideration that it deserves. David Peat maintains that physics would benefit greatly by having a description for a significant aspect of reality, specifically subtle and complex orders, such as contained in Bohm's implicate order, by creating mathematics capable of this task.[18]

Contrary to what Tegmark contends, reality is not just math. The idea held by QMT physicists that nature at its most fundamental level is only pure mathematical equations is ridiculous. Therefore, a full and accurate model of the universe will always elude mathematical representation.

Math is not the Essence of Reality

Because QMT', which is included in orthodox theoretical physics, relies so heavily on math and has high predictive power, there is enormous resistance to alternative credible explanations. These are far more intuitive and do not have QMTs absurdities in their implications and explanations for how the universe functions. However, orthodox physicists mostly ignore these, even when their predictive level is as good as QMT's for experimental results, such as from the double-slit experiment.

Assuming that math constructs are the essence of reality results in thinking that it if the math works then the theory must be correct. The section on The Myth of Quantum Entanglement discusses this further. This reveals how Bell's theorem, using just Venn diagrams and set theory, is only applicable to the imaginary world of formal logic and purely formal relations.

Outside of the abstraction of mathematics, in the real world there is no such quantity as zero for anything. It is just a concept without any instantiation. The same is true for negative and irrational numbers. Mathematics has numbers identified as imaginary, which are complex numbers with a real part equal to 0, that is a number in the bi form where b is a real number and i is $\sqrt{-1}$. The fact that all of these numbers are essential in writing math equations used in making real-world predictions does not make them real or actual in themselves. They are merely ways of referring to real things.

A huge problem with the absolute belief in mathematics is how this results in ignoring absurdities obscured by the math. An excellent example is the concept of light being both a particle and a wave based on the Copenhagen interpretation of the double-slit experiment. Werner Heisenberg, an originator of quantum mechanics, wrote about this in *Physics and Philosophy, the Revolution in Modern Science*: "The paradoxes of the dualism between wave picture and particle picture were not solved; they were hidden somehow in the mathematical scheme."[19] What Heisenberg is clearly pointing out in referring to "paradoxes" is that in spite of the contradictions and absurdities of the Copenhagen interpretation of QMT it was possible to hide these within a workable internal mathematical description.[20]

Mathematics allows for combining properties of particles and properties of waves. By doing this it is feasible to produce solutions that are compatible with particles or waves, depending on when either one is required.[21] The problem arises because these solutions are considered to have significance beyond being simply internal properties of a mathematical formalism. As a result of believing mathematics is always representative of the real world, the irrational wave-particle hypothesis derives undeserved support from the mathematics and uses this to claim the wave-particle duality theory is valid.

In the debate between idealism and realism, rather than admit that mathematics has limitations in its relevance to the real world, idealists use it, even when it's clearly not applicable, to support their absurd contentions regarding quantum mechanics. The next chapter discusses some of the absurd abstractions in QMT, in addition to those of relativity theory.

References

[1] Einstein, A. Geometry and Experience.

[2] *Ellis Ian, Science Quotes by Sir James Jeans*.

[3] Feynman, The Character of Physical Law.

[4] Bryant, correspondence with me.

[5] Tegmark, *Our Mathematical Universe*,

[6] Marmet, *Absurdities in Modern Physics: A Solution Or: A Rational Interpretation of Modern Physics.* ch 1-3.

[7] Shaw, W. *Reconsidering Maxwell's Aether.*

[8] Ibid.

[9] Peat, *Mathematics and the Language of Nature.*

[10] Ibid.

[11] Ibid.

[12] Einstein, *Geometry and Experience.*

[13] Borchardt, *matter-motion-terms-in-physics,* The scientific worldview.blogspot.ca.

[14] Ibid.

[15] *Marmet, P. Absurdities in Modern Physics: A Solution Or: A Rational Interpretation of Modern Physics. Ch 1-2.*

[16] Peat, *Mathematics and the Language of Nature.*

[17] Ibid.

[18] Ibid.

[19] Heisenberg. *Physics and Philosophy, the Revolution in Modern Science, 43.*

[20] *Marmet, Paul Absurdities in Modern Physics: A Solution Or: A Rational Interpretation of Modern Physics. Ch 1-2.*

[21] Ibid.

Part Three **Invalid Abstractions**

Chapter 4

Big Bang Theory Falsified

Dozens of BBT Problems

Preposterous Indeterminism

When Big Bang theorists modified their theory to be a zero-energy universe, this change did not solve the conservation problem. I explain the reasons in this chapter. In addition, the falsifying of the BBT demands that anyone who accepts the scientific approach abandon this discredited theory.

This chapter provides more than enough evidence to invalidate the indeterministic Big Bang Theory (BBT). The religiously inspired, highly speculative cosmogony theory that uses non-scientific foundational presuppositions has not made any significant confirmed prediction.

Big Bang advocates, envision the theory as an explosion of space and time, rather than an explosion into space and time Astrophysicist John Gribbin, who supports the BBT, stated: "It was not just matter that was created in the big bang, but spacetime as well. There was nothing outside for the big bang to explode into."[1] This last statement is not made based on some observation, but rather because it is needed to support BBT. Its proponents do this frequently to sustain their theory.

In Michio Kaku's 1994 *Hyperspace*, he states that we are discovering more evidence from experiments each year that tells us the Big Bang event occurred 15 billion to 20 billion years ago.[2] Today BBT theorists inform us that this event was 13.979 billion years ago. That is a reduction of almost one-third from Kaku's high estimate. I wonder what led to this reduction. Did some "evidence" cease to be valid?

Kaku admits that scientists do not know what occurred before the Big

Bang, and speculates that because Einstein's equations do not apply at the 10^{-33}, distances, and at the great energies involved in the origin, we need a universe with 10 dimensions.[3] He suggests that this cosmogony event began in a collapsing 10-dimensional universe becoming four or six dimensions, causing the destruction of former symmetries.[4] Kaku states that when the Big Bang happened quantum effects exceeded gravitational influences, thus we need a quantum theory of gravity, and recommends a 10-dimensional superstring theory.[5] However, because BBT is invalid, there is no need for such a theory.

Some later versions of the ever-changing BBT do not refer to an explosion, but rather an "expansion" from an extremely dense and hot state that continues today.[6] An image put forth by proponents is one of galactic dust motes carried along on an expanding soap bubble film. Representing all space and matter, the bubble expands into a mysterious primordial void, which has neither matter, time, space nor any dimensions.[7] This cosmological expansion, which is essential to BBT, supposedly results from the imagined constant creation of additional space between galactic clusters, and is self-contained within four-dimensional spacetime. Although BBT proponents believe that this is a reasonable explanation for what is occurring in a self-contained expansion, their opinion is not supported, because there is no possible way to point to what this is referencing. The whole reason for explanations is to facilitate understanding, but this one just leaves one completely confused about the expansion abstraction.

BBT uses the idea of expansion in a way designed to conform to the cosmological principle (CP), which states: "Viewed on a sufficiently large scale, the properties of the universe are the same for all observers." The CP is supposed to be true at scales of at least 250 million light years. This principle means that the universe has uniformity in all directions and homogeneity at every point due to uniformity in the action of forces anywhere in the universe. However, several recent discoveries show this principle to be incorrect.

One such significant discovery is the Hercules–Corona Borealis Great Wall found in November 2013[8] 10 billion light years away. It measures 2000-3000 Mpc, which is between 6,520,000,000 and 9,780,000,000 light years.[9] That is 39.12 times larger than the size where homogeneity is supposed to occur. This discovery alone invalidates the cosmological principle; at least at the scale that the CP is stated to be applicable.

Glenn Borchardt addresses the question of whether one can falsify the

CP. Here is an excerpt from a message that he sent to me on June 25, 2016, regarding this issue: "I have never read a clear statement of the CP that would allow it to be falsified. One is that, at a large enough scale, the universe looks the same from all locations; one is that the universe is homogeneous. Both are idealistic and impossible because identities are not possible. The large enough scale is the clinker. No real portion of the universe ever follows the CP. When that fails, as it always does, they just jack up the size requirement."[10]

Based on BBT supporters' history of continually increasing the scale at which the cosmological principle applies, I would not be surprised to see them enlarge it to a size greater than the observable universe. That creates a serious problem for the theory because there would not even be a theoretical possibility of ever falsifying it.

I disagree with philosopher of science Karl Popper in his use of falsifiability as the demarcation criterion for discerning science from non-science, because a theory could have much value, but one may not have a way to falsify it. For example, it is not possible to falsify the Infinite Universe Theory (IUT), which has many proofs and is without contradictions. This is not a related situation as exists with the CP. The significant difference between the two theories is that whenever discoveries of new massive structures in the universe falsify it, CP's size requirements change, whereas this does not and cannot occur in IUT.

Despite my disagreement with Popper on this, nevertheless, I completely support the idea that being able to falsify a theory can add to its credibility. Any theory that frequently needs to modify its stated conditions when observations contradict it is not viable, and we need to abandon it rather than try to find ways to preserve it. This situation applies to the BBT, the Inflation Theory, and the CP.

BBT needs CP, but this principle is not valid. This means the BBT is using an invalid principle for support, thereby making BBT an invalid theory. Unlike BBT, the Infinite Universe Theory discussed in Chapter 9 does not require the CP.

I suppose I should now be referring to a big expansion and inflation theory, but it is still widely known as the Big Bang theory. Therefore, I will use that term for this invalid theory.

BBT and relativity theorists are adamant that space is not really a void, because it contains vacuum energy, which is the underlying background energy everywhere in space; some of this energy may be from virtual particle pairs that supposedly come into existence then annihilate each other immediately. What are termed virtual particles are mislabeled, because they are actually real, having established properties and consequences, with effects that are observed and measured by physicists.[11] Heisenberg's energy-time uncertainty principle codifies this behavior.

Resonances in the space between metal plates close together supposedly creates an attractive force known as the Casimir effect identified as vacuum energy. Physicists believe that vacuum energy "exists" in spontaneous emission and in the Lamb shift, which is an energy difference between two energy levels of a hydrogen atom.

Vacuum energy supposedly has cosmological scale affects.[12] Astrophysicists estimate free space's vacuum energy at between 10^{-9} joules and 10,113 joules per cubic meter.[13] This theory also posits that the vacuum implicitly has properties identical to particles, such as energy, polarization, and spin, etc., most of which cancel out on average, leaving only the "vacuum expectation value" or vacuum energy. Because of this, physicists consider vacuum energy just as real as magnetic fields or electrons etc.

While thinking about space filled with vacuum energy, consider the following explanation from BBT theorists. They posit that when the Cosmos was only a year old, during the rapid period of inflation, the separations between galaxies were increasing at 300 times light speed.[14] Unbelievably and shamelessly, they make the ridiculous claim that space, which they envisage filled with vacuum energy, can expand faster than the speed of light because space does not represent matter or energy. That has to be one of the most blatant contradictions. In the novel *1984*, Eric Blair (a.k.a George Orwell) termed this "doublethink."

Attraction

The idea of gravitation as an attractive force comes from Newton's law of universal gravitation, which states: "Every object in the universe attracts every other object with a force directed along a line of centers for the two objects that is proportional to the product of their masses, and inversely proportional to the square of the separation between them."[15] As referred to in Chapter 11, by 1718 Newton no longer believed that gravity was an attractive force.

In the text *Physics Principles and Problems,* appears the statement:

"The question of why objects attract each other is still not answered."[16] No one has answered this, because it is a wrong question.

Attractive force has no part in gravitation in general relativity, which contends that it results from the curvature of spacetime by masses. Einstein's equivalence principle means that gravitation is the same phenomenon as inertia. This is a property of matter resulting in stationary objects remaining as such, and moving objects maintaining their speed and direction until affected by something.[17]

Einstein decided that the universe was unbounded, yet finite. Big Bang theorists used this contradictory idea to support their model. Thus, the foundation for BBT is a contradiction.

Curved Space

In imaginary four-dimensional GRT, no central point exists. A finite unbounded universe with curved space needed no macrocosm in expanding into itself. GRT used the concept of curved space around all objects. The relationship between any two bodies meant that no curves were perfect, because each has its own curved space that results in an "attractive force," or gravitational field.[18] Generalizing this curved space abstraction to the greatest extent resulted in the idea of a curved universe. Additionally, this four-dimensional curved universe required thinking that time is matter, rather than motion. The primary difficulty with GRT is the impossibility of conceiving of it. Because BBT uses the same model, it shares the same problem.

Chaos and Heat Death

The chaos concept envisages the end of all matter based on the SLT, which states that as entropy increases, isolated systems become increasingly chaotic. Because the BBT assumes that the universe is an isolated, finite system, it projects the same fate for it. However, the universe is neither isolated nor finite. The chaos idea comes from the Bible, and BBT's origin has a religious influence. Georges Lemaître, an astronomer and Belgian priest, proposed this theory of the universe's origin, which he called his *"hypothesis of the 'primeval atom' or the Cosmic Egg."* Hannes Alfvén stated that Lemaître revealed to him that *the Big Bang Theory was designed to reconcile science with St. Thomas Aquinas' religious dictum of creation out of nothing or creatio ex nihilo.*[19] The goal appears to be to demonstrate that compatibility can exist between science and religion.[20]

By generalizing the SLT's increasing disorder results in the concept of the "heat death" of the universe, with heat transmitting to low heat areas from high ones until equalization of temperature exists throughout the universe. However, this all depends on going from the idea of perfectly isolated systems, which in reality never existed, to an imaginary completely isolated universe, which is impossible. Heat death requires the absurd abstraction of motion having separateness from matter and disappearing into a void. As motion does not refer to a thing, but rather to what matter does, motion cannot vanish in emptiness. Chaos does not refer to the universe but rather to undisciplined thinking, some, something BBT proponents excel at.

Cosmic Expansion: Gravitational Redshift

Redshift is a spectral shift in visible light's color from a galaxy when any electromagnetic radiation (EMR) including light, increases its wavelength to the red end of the spectrum equivalent to reduced photon energy and frequency. Orthodox cosmologists' interpretation of this phenomenon is that it results from the "Doppler effect," gravitational effects and cosmic expansion, which is erroneously presumed.[21]

Redshift refers to incremental increases in EMR wavelengths at related distances. There is zero redshift for any radiating atom or molecule on Earth, which is stated as $z=0$. If an object from another galaxy has a wavelength of 2, which is double that of standard wave lengths on Earth, then its redshift is 1 (i.e. $z=1$). The B factor in the Hubble formula converts redshifts to distances, although redshifts are not proportional to distances. For example, a redshift of .09 equates to approximately 1.169 billion light years. A redshift of 7 is equal to about 13.172 billion light years, which according to BBT approaches the outer areas of the observable universe.

Astrophysicists who support the BBT use an indeterminist interpretation of redshift, considering it the result of the expansion of space, attributed to the effects of the Big Bang.

Without resorting to expansion, Halton C. Arp accounted for redshift in a paper published in the *Journal of Astrophysics and Astronomy* in September 1987. He found that the continuity of redshifts less than or equivalent to 1,000 km per second of the observed galaxies, and isolation of these concentrations, mean that their distance equals the brightest members of the group. The fainter ones have higher redshifts, which mimic to some degree, a Hubble relation. Arp concluded that the higher redshifts must be due to a cause other than velocity, because the galaxies are all at the same average distance.[22] BBT true believers dismiss any non-Doppler effect explanations of redshift, without actually addressing the strong observational evidence.

The Cambridge University Press 2011 reissue of Arp's 1987 book titled *Quasar, Redshifts and Controversies* provides much information on his theory and his observational results.

Redshift, Dark Matter and Molecular Hydrogen

Large amounts of hard to detect molecular hydrogen (H_2) in space accounts for unusual galactic rotational motion, and for slowing light wave propagation through interstellar space. These effects result from hydrogen collisions[23] with aether, which is the wave medium for light causing an observed redshift.

A shift in frequency of a wave caused by the relative motion of an observer and an emitting source, means a Doppler shift has occurred. BBT advocates claim this phenomenon accounts for the observed frequency downshift toward the red end in spectral lines. This is supposedly because these far away galaxies providing the light source are moving away from us, which would signify an expanding Cosmos.

In 1999 Dr. Edwin E.A. Valentijn and Dr. P. van der, Werf observed molecular hydrogen in enormous quantity in an edge-on galaxy in Andromeda, 30 million light-years distant. In the *Astrophysical Journal Letters,* of September 1, 1999, they wrote that it "matches well, the mass required to solve the problem of the missing mass of spiral galaxies."[24] They state: "It is well established that if there is about 10 times as much molecular hydrogen as atomic hydrogen in the disks of spiral galaxies, then the missing mass problem [in galaxies] is solved."[25] In 1999, detection occurred of molecular hydrogen absorption lines in the important far ultraviolet (FUV) spectral range in a high-velocity galactic halo cloud of the Milky Way.[26]

Atomic hydrogen is the simplest stable atom. Because it can change its spin, thereby changing its energy, it is extremely easy to find in galaxies and intergalactic space with a high frequency radio signal at 21-cm wavelength. Looking for H_2 in space is not common and it is very difficult to detect. This is because electromagnetic radiation emits at a wavelength of 21-cm, but when two H atoms combine into H2, their coupled spins and electric field cancels out, thereby eliminating the radio-frequency spectral line at 21-cm. Thus, H2 is not visible at that wavelength.

Unlike most molecules composed of two atoms, H2 has no dipole, which refers to a permanent pair of equal and opposite electric charges or opposite magnetic poles a small distance apart. So light going through it has no electric dipole to become excited that would emit or absorb energy. This renders detection of EMR from it virtually impossible. The huge amount of H2 very significantly undermines the credibility of astrophysicists' statements about knowing the amount of hydrogen in the Cosmos.[27]

Based on vast quantizes of extremely stable H in the universe , the relative abundance of H compared to H_2 in space, and that much H_2 was naturally transformed into H, we can logically anticipate many more H_2 discoveries.[28] This huge amount of transparent H_2 in space interacting with light is enormously significant for interpreting cosmological redshift and thus important consequences for the hypothesized origin of the universe by removing its principal support for an expanding universe.[29]

When matter recedes from us, the light frequencies it emits shifts towards lower frequencies or the red end of the spectrum. Interpreting cosmological redshift using the ideas in GRT, results in the explanation that it occurs because spacetime's curvature was less when the universe was younger than it is now. Light waves stretch en route between their emission and detection. However, for this to be valid, inconceivable Minkowski spacetime would have to be other than pure imagination, which is impossible because time is motion, not a dimension.

Physicists describe light as a "wave packet" or "wave train", which is a brief burst of localized wave action moving as a unit. One of its components is low frequency, which depends on the duration of coherence of the EMR, and the other is the usual high frequency. Because of its coherence maintained during a specified period called the "time of coherence," there is regular progression of the phase of the electromagnetic field over time.[30]

Aether, which is the wave medium for light, endures an incredible number of collisions with H_2. Due to these collisions, some light wave energy is lost in scattering depending on duration of coherence. Because of aether's small degree of inelasticity when colliding with H or H_2, the ensuing redshift appears identical to the cosmological Doppler interpretation of redshift.[31] This produces the appearance that galaxies are receding as an effect of the Big Bang.

Present technology detects H_2 that has the highest rotation quantum numbers and vibration, which is the warmest H_2. When we can find the coldest H_2, the amount may very likely equal the missing mass of spiral galaxies.

Non-expansion Velocity Causes of Redshift

BBT does not explain the actual mechanism causing many redshift observations. It attempts to account for cosmological redshift by the expansion of space between galactic clusters, but this fails to explain the following phenomena:
1. CMB radiation uniformity.
2. The "Fingers of God" effect of elongated galaxy distribution in redshift space, with elongation axis toward the observer.
3. A single redshift galaxy in compact galactic groups.
4. Solar limb redshift.
5. Quasar's intrinsic redshifts.
6. "K-effect" correction to a cosmic object's magnitude or flux (total energy crossing a unit area per unit time (i.e. nadal) as joules/square metre/second) that enables measuring light at a redshift converted to equivalent measurement in the object's rest frame.
7. Redshift.

In a 2016 paper, Louis Marmet discussed 20 physical redshift mechanisms not based on expansion velocity.[32] If a mechanism can somehow cause light to lose energy, this will cause redshifts, but to agree with observations it must not cause light scattering. Borchardt writes: "It would be impossible for light to travel billions of light years without losses."[33] The solution could be flowing aether particles, which being far smaller than quantum entities do not scatter light, but would cause light to redshift, as energy is lost from moving through it. Galactic redshift would be proportionate to distance.

Aether as cause of Redshift

In Borchardt and Puetz's Neomechanical Gravitational Theory,[34] light's velocity is highest where aether density is greatest, which is between galaxies. We see galactic redshift proportionate with how far light travels in those areas because every light wave occurs over a greater distance than in our solar system.

For that portion of light's journey through ordinary matter areas, such as our galaxy and solar system, we observe far less redshift. Borchardt writes: "Aethereal Redshift Theory (ART) states that the density of the aether medium and thus the velocity of light increases with distance from baryonic (ordinary) matter. For a particular light frequency, an increase in velocity produces an increase in wavelength, which appears as the misnamed "gravitational" redshift."[35] Aethereal redshift is an inverse function of baryonic matter density because light velocity relates to aether density. This explains redshifting of light from massive objects, decreasingly affected by ordinary matter held by gravity.

Pound and Rebka[36] showed blueshifting of EMR waves coming toward Earth's surface, where less compacted aether affects the light, and redshifting for outward-bound waves. Thus, aethereal redshift is the mislabeled gravitational redshift.

Explanation of CMB

CMB is a wide spectrum electromagnetic radiation (EMR) having a bell-shape as given by the Planck equation, which depicts EMR coming from any cosmic body of any temperature. BBT assumes that CMB originated from the Big Bang when it was transparent at 3000°K. Because of its extremely high expanding velocity, there is a redshift of that Planck radiation at 3,000°K. In considering the supposed Doppler effect, cosmologists measure CMB at 3° K.[37]

Any surface above absolute zero emits the Planck spectrum. Some spectral lines matching the quantum states of atoms that are electronically excited above the surface appear superimposed to the Planck spectrum, such as plasma above the Sun's surface. By measuring the Planck spectrum in the infrared, one can obtain a cloud's temperature, which is usually minus 70° Celsius. BBT proponents attribute the 3° K cosmic radiation to the Big Bang, but it is more likely from enormous amounts of H_2 emitting heat throughout the Cosmos, which Gerhard Herzberg, winner of Nobel Prize in chemistry, measured at this same temperature.[38] Thus, it is clear that the Planck spectrum originates with emissions from H_2. As only one Planck spectrum appears to observation, it cannot be from the Big Bang. With the loss of its principle evidence–the CMB temperature–the BBT is not tenable.

The evidence provided by BBT cosmologists does not support it. The rest of this chapter presents some of the ever-increasing evidence that contradicts BBT, which is far more than is needed to invalidate this impossible idea.

Cosmic Inflation Concept and Dark Matter

Alan Guth, the originator of the inflationary theory, writes in *The Inflationary Universe*: "Despite its name, the big bang theory is not really a theory of a bang at all. It is only a theory of the aftermath of a bang."[39] He points out that the BBT's equations depict the way the primordial fireball expanded and then cooled to form the structures of the universe without referring to what precipitated the bang or what banged, something that inflation attempts to explain.[40]

Inflationary cosmology is not a single theory, but a cosmological framework constructed on the concept of repulsive gravity causing expansion of space. According to Guth, inflation refers to a rapid cosmic expansion, in which gravity was a powerful repulsive force; increasing the volume of the universe by at least a factor of $10.^{78}$ Expansion supposedly emerged from a realm that was "wild, chaotic, energetic."[41] Inflationary cosmology asserts that the Big Bang occurred once the required conditions were present. These included having an inflation field for available energy and negative pressure needed for repulsive gravity. The theory asserts that the already existing universe experienced an inflationary bang, which was not necessarily the creation event of the universe.[42] That is an ambiguous and unclear idea, and it shows confused thinking. If the universe already existed, then the inflationary bang did not create it. However, the phrase "not necessarily" implies that it could also have been the creation event.

Inflationary theorists conjecture that the inflationary epoch started 10^{-36} seconds post Big Bang and ended at between 10^{-32} seconds to 10^{-36} seconds. Guth proposes that a seed tinier than a proton, which included all mass-energy that exists today, inflated to the size of a basketball within this incredibly brief duration.[43] He does not explain what caused inflation to occur. During this rapid inflation a smoothing out took place for any irregularities in matter's distribution and even in space's structure.[44] It is hypothesized that at inflation's conclusion the expansion energy transformed into mass, energy and matter, which combined with the newly generated protons and electrons. Guth does not appear to understand that energy, which is a matter-motion calculation, does not transform into matter. That is logically impossible. According to this theory, the diminishing of the repulsive gravity of inflation enabled the so-called "attractive" gravity to dominate, which resulted in the clustering of matter into galaxies and stars, etc.[45] According to this theory, 7 billion years post bang, on a cosmic scale the size of galactic clusters, repulsive gravity once more became dominant for reasons unknown.

Inflation theory proposes that because this additional dark matter lacked any electric charge, radiation could not make it smooth, enabling the dark matter to cluster together causing irregularities in this expanding universe.[46] When decoupling of radiation and matter occurred, dark matter was 10 to 100 times as common as regular matter. This inflation with the dark matter scenario, supposedly accounts for the smooth CMB by the time of decoupling.

Since Guth first announced his inflation theory in January 1980, cosmologists still have not decided on what caused the hypothetical inflation. They only have speculations such as the strong nuclear force separating from the other elementary forces at the time, thereby producing a positive vacuum pressure or negative vacuum energy density. They further surmise the occurrence of a type of phase transition, such as water becoming ice, which left the universe very unstable.

The vacuum pressure or energy resulted in a marked anti-gravitational effect that acted as a strongly repulsive force, stretching the boundaries of the universe from the size of a proton to that of a grapefruit.[47] I thought inflation proponents had agreed it was the size of a basketball, now they say that it is grapefruit-sized.

There are two significant issues with this theory. First, based on a 500 million light-years diameter of the universe at 380,000 years, this means that during inflation the universe expanded faster than the speed of light. Because I use the concept that space is matter, I do not accept that this speed is possible. The inflation theorists argue that it is possible, because they do not consider space to be matter. On the one hand, BBT posits that space is a void that is inflating, thereby causing the galactic clusters to move apart. They do not offer a reasonable explanation of how a void is capable of expanding. Contradictorily, they also claim that this void contains fields. Although I realize the necessity for the BBT model to include this concept in attempting (unsuccessfully) to explain how expansion can occur at such absurdly fast speeds, I cannot understand how anyone can accept such a ludicrous idea. It sounds like something from the world of *Through the Looking Glass, and What Alice Found There*, by Lewis Carroll. There is also the issue of why a law only operated for this incredibly tiny duration.

In *The Fabric of the Cosmos*, physicist Brian Greene writes that we have no idea of what conditions existed prior to inflation and the probability for its occurrence is unknown.[48] As Greene admits, the reason for believing in inflation theory is that it offers a solution to apparently separate problems: the origin of structure, low entropy of the early universe, flatness, and horizon.[49]

Dr. Paul Steinhardt, a cosmologist and theoretical physicist at Princeton University, co-authored the underpinning papers in the 1980s for the inflationary theory. He is now one of its severest critics, because it produces a multiverse, creating infinite kinds of patches of space for any possible cosmological result. These limitless patches are not homogeneous, isotropic, or flat, and lack nearly scale-invariant perturbations.

Steinhardt points out that by producing the multiverse, the inflation hypothesis has so much flexibility that any number of observations cannot falsify it, and no observations since its creation support it because the theory accommodates any result equally.[50] Considering that there is no possibility of disproving it, inflation theory does not qualify as science.

Steinhardt emphasizes that inflation is not a solution to the horizon problem. He comments that it is an improbable accident for the universe to have the isotropy and homogeneity, which apparently occurs.

Planck satellite observations from 2013 and 2015 along with previous ones from SPT, ACT and WMAP, and other experiments, eliminate a wide spectrum of complex inflationary models in favor of single scalar field models.[51] Significantly, every simple inflation model lost favor compared to ones with plateau-like potentials. Remaining models needed fine-tuned additional specifications combined with further unlikely beginning conditions than required by the simplest models. Cosmologists term this situation "the unlikeliest problem."

In their 2013 paper, Iljas, Steinhardt and Loeb [52] comment that until recent data from WMAP, ACT, and Planck 2013, classic inflation's main problems were conceptual: "multiverse unpredictability," the "Liouville problem," and the "entropy problem." They conclude that the data shows that even if inflation cosmologists find solutions for all the conceptual problems, observations are disfavoring classic inflation.

Universe is Not Expanding

The expansion of the Cosmos is the most basic hypothesis of BBT, but observational facts contradict this assertion, which necessitates constant add-ons and adjustments to this deeply flawed theory. For example, Einstein created the concept of the repulsive cosmological constant force to explain why the Cosmos did not succumb to its own immense gravitational influence and contract. With the later discovery of galactic redshift, supposedly resulting from the stretching of galaxies away from us with expanding space, Einstein dropped his idea because now this sufficed as the reason that cosmic collapse does not happen.

In 1922, mathematician/cosmologist Alexander Friedmann developed a solution to Einstein's field equations of General Relativity. It required an expanding cosmos, which became the foundation of the BBT, although it also was a basis for steady state descriptions. Using three different models, he explained how observations of an expanding cosmos would occur everywhere with an isotropic and homogenous appearance. He did this for each of three different models.

In an open universe where the amount of universal expansion outstrips the escape velocity of all matter, space curves negatively, and expansion never ends. A closed universe has positive curvature, with gravitation ultimately ending expansion and causing it to contract back to its size at the hypothesized Big Bang. It is theoretically possible that in this model, the universe could have endlessly repeated expansions and contractions. The flat universe model balances between open and closed as expansion continues endlessly until after infinite time its velocity moves toward zero with no curvature in space. (Apparently, Friedman defines "infinite time" as an extremely long duration, but not one that continues on forever.) The present consensus of BBT uses this last model, sufficient matter exists such that it eventually will halt expansion.

The basic premise of BBT, that the Cosmos is expanding, is an assumption, not a fact confirmed by observation. There are four classical tests for expansion based on the assumption of redshift of galaxy light representing the velocity of the galaxy moving away from us. The tests concern relationships between galactic redshifts. It is indisputable that distance of ordinary galaxies correlates with their redshift, but we do not have any proof that shows that increased distances causes redshift.

The first test is redshift for galaxies versus apparent magnitude. Closed universe expansion models appear to correspond with observations, which also fit with a static universe model with no free parameters. Because static models are simpler theories, the test favors these based on the application of Occam's razor, because they do not require inventing unnecessary hypotheses, and they have fewer free parameters.

The second test is redshift versus number of galaxies. Except in the case of an open universe, and with the application of evolutionary corrections required at a redshift of 0.4, a period when galaxies were similar to today's, this test does not agree with expansion models. It supports static models, which do not require more parameters.[53]

Test number three compares galactic redshift with surface brightness. Attempting to correct for the Malmquist bias effect, which causes detection of a disproportionate number of intrinsically bright objects, presents the greatest challenge of any in this test, making it difficult to draw conclusions. This biased data agrees with both models, although they have higher correspondence with expansion.[54]

The fourth test of galactic and radio sources redshift versus angular size has the least observational bias of the four tests. Observation results of galactic cluster radii, and independently from brightest members of clusters[55] support static universe models, while contradicting the predictions of all Friedmann expanding universe models.

Although results from the greatest angular sizes of double radio sources offer less support for static models, they have strong divergence with every expansion model. Absences of small radio sources in results provide cosmological parameters that are not consistent with expansion or static models. However, they could indicate interplanetary scintillation effects,[56] which are momentary, random fluctuations in celestial radio waves' intensity, similar to a star's twinkling in the visible electromagnetic spectrum.

Of these four tests, three of them support static universe models, with the fourth test not applicable to most static models. The only test that agrees with expansion has problems with observational bias.

BBT proponents need to hypothesize strong evolutionary effects, including ones that are counter-intuitive, such as the intrinsically smallest radio sources are the strongest ones. BBT requires that there be very little or no deceleration of the universe, or accelerated expansion.

Big Bang theorists describe the Cosmos as strongly open, which refers to a model in which expansion continues without end due to insufficient mass to balance it by gravitational influence. However, all observations from small radio sources imply a strongly closed universe, which is one with enough matter and therefore sufficient gravitational force to stop the expansion. For static models, we only need to assign significance to galaxy results in agreement, because redshift does not correlate to distance for radio galaxies, quasars and other radio sources.[57]

Two other tests are also available to test for expansion. One involves a test of supernova light curves. If expansion were occurring, then we would observe a stretching of these in high-redshift galaxies resulting from the originating galaxy rapidly receding. This test yields ambiguous results.[58]

Another test uses galaxy evolution. If the Big Bang was 13.799 billion years ago, the appearance of galaxies is relatively recent, and their evolution would be a significant aspect of the young Cosmos. In a non-expanding universe, present galaxies would be similar to those of 12 billion or 13 billion years ago. The radio galaxy 3C 65 at a redshift of 1.2 would be larger and fainter than any present local galaxy.[59] This necessitates evolutionary effects. However, because distance does not correlate to redshift for quasars and most radio galaxies in static models, it is obvious that we cannot make size and intrinsic brightness inferences for radio galaxies. Static models do not interpret a quasar's development to ordinary galaxies in the same way as the BBT, which rules out using quasars in any test. An additional test of expansion is the Tolman test, described in the next section.

The "Butcher–Oemler Effect" hypothesis is that a greater percentage of blue galaxies are in the cores of galaxy clusters ($z \sim 0.3$) than in the cores of low-redshift clusters. Astronomers presume this to be evidence that galaxies are constantly evolving..

There are many more faint blue galaxies at redshifts of 0.4 and higher up than appear to exist in our local area. However, it seems likely that these distant faint blue galaxies have a local counterpart in galaxies with low surface brightness (LSB).[60] Years ago, astronomers had difficulty finding LSB galaxies. However, presently astronomers find them as frequently as spiral galaxies[61] when constructing surveys specifically designed to locate them, although they are more blueshifted. The close resemblance between spiral galaxies and faint blue galaxies clearly invalidates the theory of galactic class evolution.

More Evidence for No Expansion

We must keep in mind that the theory of cosmic expansion is just conjecture without any real evidence that non-BBT models cannot also explain. Big Bang theorists invented this because they needed it to account for the universe's diameter.

Sumner has recently pointed out that introducing new space in expansion dilutes the vacuum's permittivity.[62] This results in changes to the frequency around atoms, which has twice as much effect on observed redshifts as does the speed of expansion. When the equations include this, it indicates a contracting universe.[63] This is something BBT would not be able to explain.

The Big Bang Theory depends completely on interpreting galactic redshift of light as being principally the outcome of the Doppler effect. This redshift is proportional to distance, which astrophysicists infer from brightness. However, the only explanation for redshift those astronomers considered possible was the velocity of the light source moving away from us. Therefore, their inevitable conclusion was that a galaxy moves faster as its distance from us increases. Without this connection, there would be absolutely no basis for an expanding universe or the extrapolated Big Bang.

Until the concoction of the BBT, it was an axiom that conformed to common experience that as an object gets further away from us its brightness per unit area (surface brightness) stays constant while it appears smaller and dimmer. However, BBT advocates claim that at galactic distances, an expanding Cosmos must cause these more distant objects to look larger but fainter, and the light becomes dimmer from the stretching effects of expansion. Therefore, BBT supporters postulate that with increased distances, surface brightness decreases. This means that our strongest telescopes would not be able to detect the furthest galaxies, because according to the BBT they should have surface brightnesses dimmer by hundreds of times compared with equivalent proximal galaxies.

However, a study by astrophysicists Eric J. Lerner, Renato Falomo, and Riccardo Scarpa[64] from Lawrenceville Plasma Physics contradicts the BBT prediction that Euclidean geometry does not work at great distances. Their study, published May 2, 2014, in the *International Journal of Modern Physics D* provides strong support for the concept of a non-expanding universe.

Big Bang Theory Falsified

Picking about 1,000 of the most luminous spiral galaxies, they matched close and extremely far away ones for average luminosity, and compared their size and brightness. Their observations showed identical surface brightness of galaxies, irrespective of their distances. This result corresponds to expectations from Euclidean geometry in a non-expanding Cosmos, thereby refuting the BBT expanding universe model in which surface brightness diminishes with increased distance. The researchers applied the Tolman test, devised in 1930 as a test for determining whether expansion was occurring by using dimming of surface brightness as the criteria. After analyzing UV surface brightness of luminous disk galaxies from GALEX and HUDF datasets, reaching from the local Universe to $z \sim 5$, they concluded that surface brightness stays constant as would be expected in a Static Euclidean Universe (SEU).[65]

Lerner commented on the results: "Of course, you can hypothesize that galaxies were much smaller, and thus had hundreds of times greater intrinsic surface brightness in the past, and that, just by coincidence, the Big Bang dimming exactly cancels that greater brightness at all distances to produce the illusion of a constant brightness, but that would be a very big coincidence."[66]

There was another surprising result to their study. In order to apply the surface brightness test, and match far and close galaxies, they first needed to determine each galaxy's actual luminosity by linking redshifts with distances. Extrapolating the well-substantiated observation that redshift remains proportional with distances in the nearby universe, they applied this to all distances. The scientists checked this relation between distance and data on redshift supernova brightness used in determining the supposed accelerated expansion of the universe.

Dr. Renato Falomo of Italy's Osservatorio Astronomico di Padova commented: "It is amazing that the predictions of this simple formula are as good as the predictions of the expanding Universe theory, which include complex corrections for hypothetical dark matter and dark energy.[67] Study co-author Dr Riccardo Scarpa from the Instituto de Astrofisica de Canarias in Spain, commented: "Again you could take this to be merely coincidental, but it would be a second big coincidence."[68]

The researchers concluded that without expansion as a factor, redshift with greater distances has to be resulting from something that happens to light on its journey here through space. Lerner stated: "We are not speculating now as to what could cause the redshift of light, however, such a redshift, which is not associated with expansion, could be observed with suitable spacecraft within our own solar system in the future."[69]

Is it possible that many scientists oppose the abundant and strong evidence presented in this chapter regarding other causes of galactic redshift because they have linked their careers to the Big Bang model and they fear the strong sanctions that apply to dissident scientists? I think it is very likely that this applies to most BBT cosmologists.. If irrefutable evidence emerges for alternative explanations for galactic redshift, then BBT will lose most of its supporters.

Tired Light Theory (TLT)

This is the best candidate for redshift not being completely the result of the Doppler effect. This theory involves light experiencing losses due to absorption while travelling billions of kilometers, because it is impossible for anything to travel those vast distances without experiencing any losses. These losses may be sufficient to account for part of galactic redshift.

In comparing the no-evolution, $q0 = 0$, expanding universe model with the no-evolution, tired-light model, four types of observational tests produced data that better conforms to a tired light hypothesis.[70] Additionally, the latter do not require ad hoc assumptions concerning rapid galactic evolution.

The simplest interpretation of results requires assuming significant evolutionary effects, Euclidean space, cosmologically static galaxies and photon energy loss of about 6 percent over 10 billion light years. This stepwise incremental energy reduction explains redshift quantization observations.

In expansion models, redshift quantization requires multiple assumptions, but this is not the case in the simpler tired light model. Additionally, this model has the advantage of conforming to observations of a slowly evolving, stationary Euclidean universe with gradual photon energy loss over cosmological distances.[71]

Big Bang Singularity

The hypothesis that the universe expanded from a singularity creates a massive problem for BBT by violating two assumptions of science. It contravenes the infinite time aspect of infinity by having a beginning, and breaches infinity by being a closed system in a singularity with nothing beyond it, leaving the universe without an opportunity to interact with a macrocosm.[72] It also is problematic because mathematics indicates that the laws of physics are not applicable at a singularity (Davies 2004).[73] The singularity concept also contradicts the assumption of inseparability.[74]

Hannes Alfvén

Plasma physicist Hannes Alfvén recipient of the 1970 Nobel Physics prize, referred to the cosmological pendulum idea: i.e. that cosmology alternates between myth and science as a perspective. Creationist and scientific cosmology worldviews have been recurrent. Today's' myth of mathematical perfection was also prevalent in the times of Plato and Ptolemy. For Alfvén the cosmologists' obsession with mathematical perfection is the mythical approach's foundation. In Alfvén's view, the difference between what is a myth and what is science is that the latter require theories developed from real-world observations, whereas the former emerges from inspiration of unaided reason[75]

Alfvén regarded BBT as pseudoscientific fantasy to explain creation, seeing no rational reason to think that the universe had a beginning[76] and that all cosmogony theories are myths of how the universe came to exist.[77] Alfvén contended that the basis for the BBT is ideology rather than the scientific method and that when people consider the universe, a conflict always arises between the empirical and mythical approach.[78] In the latter, one seeks to explain what perfect principles were employed by gods in creating the universe.[79] He reasoned that because religion does not accept empirical methods, it is unwise to attempt to reconcile religious belief with scientific theories.[80] Alfvén pointed out that although an evolving universe without a beginning does not conform to the Biblical story, other religions, such as Buddhism, do not require an origin of the universe, and that the Christian doctrine of *creatio ex nihilo* did not even exist prior to AD 200.[81] In his estimation, the important thing is not to confuse religious fabrication and scientific results.[82]

Alfvén explained the flaw in the BBT's manner of construction. Its creators did not begin its formation from observable phenomena, but rather attempted to use mathematical theories to make a deduction of the cosmogony of the Cosmos. From that point, they project to the present. He shows how this is similar to that of British-American geophysicist

Sydney Chapman's theory that currents could flow only in the ionosphere with no downflowing currents. It was widely accepted for decades because of its mathematical elegance until satellites measuring downflowing currents falsified it in 1974 [83]

Similar to Chapman's approach, Alfvén contends that BBT's fundamental error, which has resulted in multiple observational contradictions, was trying to predict today's Cosmos from an imagined beginning to the universe. In contrast with that method, Alfvén's approach arrives at theories through extrapolating from observations in the laboratory of space probes. It is necessary to use recent events in any description of previous times in the history of the Cosmos. Because we do not observe matter in the process of being created from nothing, the scientific assumption is that this never happened and the infinite Cosmos has neither beginning nor ending.[84]

Although there are dozens of serious problems, any one of which could invalidate BBT, its proponents only focus on three of these, which are: "horizon," "flatness" and "monopoles" problems.

The Horizon or Isotropy Problem

A cosmological horizon is a measurement of the furthest travelling distance possible for information, such as heat energy, due to being limited to the speed of light. The horizon problem refers to extremely distant regions of the universe having very similar temperatures, and to the CMB reaching us from all directions being almost uniform in all directions (isotropic). After the Big Bang, these regions were never near enough to one another to have reached the same temperatures through energy exchanges.

Origins of CMB include regions for the Cosmos that were expanding in opposite directions. This indicates that all regions of space since the thermal radiation was emitted have always been too separated, and thus beyond the other's horizon to ever have exchanged heat even at light speed. Yet their temperature is within 0.01 percent of CMB anywhere else. When the photons left their source, they were 100 times further apart than any connection at the speed of light

BBT proponents attempt to explain the problem by fabricating the implausible and unsubstantiated idea of inflation, which contends that a process of exponential and incredibly rapid expansion of space occurred for the early universe, lasting from 10^{-36} seconds to 10^{-32} seconds following the Big Bang. This increased the size of the universe by a factor of somewhere between 10^{20} to Alan Guth's estimate of 10^{78}, which involved expanding at 116 times light speed.

Their proposed horizon problem solution contends that inflationary expansion was so rapid that homogeneity did not have time to break up. Therefore, following inflation the universe was homogenous and isotropic even though its parts had separated too much for any energy exchanges. This explanation requires that we accept the preposterous idea that the inflationary expansion occurred at such incredible speeds, and that it happened despite having nothing into which to expand.

They maintain that our observable universe resides "inside that expansion" and isotropic radiation results from all space inflating from a tiny volume that had identical initial conditions.[85] Whoever concocted this response does not have a good understanding of logic and language. Stating that the Universe is "inside" of the process of inflation that ended almost 13.8 billion years ago is ridiculous. Can anyone be inside of their growth process? Perhaps the individual who thought up this impossible solution is meaning that the universe is inside of itself. However, that is illogical and absurd. If this is the best answer that the BBT true believers can come up with, they would be wiser to simply not answer.

Dr. Paul Steinhardt, referred to in the "Cosmic Inflation Concept" section, emphasizes that inflation is not a solution to the horizon problem, because it results in a multiverse with infinite patches that are neither isotropic nor homogeneous. He points out that in the multiverse mode it is an improbable accident for our universe to have the isotropy and homogeneity, as it appears to have.

Another problem with inflation theory is the inflation field's potential energy curve, seemingly devised to accommodate any possible data. Additionally, no recognized physical field corresponds to the hypothesized inflation field.

The flatness or Oldness Problem concerns the shape of space and involves a cosmological density of matter and calculation representing energy. The problem arises because the Universe's original condition appears fine-tuned to very specific values, for which even tiny divergences would affect its appearance today in an extreme

Thus, at the moment of the inflationary bang, if the total matter and energy per volume unit of space differed by the tiniest bit from the critical density, then expansion would result in the universe going way beyond critical density. A very specific value is required for the necessary flat universe. Today's Cosmos has a density very close to this critical value.

Over vast cosmological time, density deviates rapidly from the critical value, which significantly increases the flatness problem. Therefore, the new Universe's density must have been within one part in 10^{62} or less of critical density. This leads cosmologists to wonder what factors resulted in the initial density that fine-tunes to this "special" value. BBT supporters have no answer to what factors caused the original density's fine-tuning to such a precise special value.

Physicist Rod Nave in referring to the total amount of matter that exists stated that the Universe might be very nearly flat, or barely closed or barely open.[86] He illustrates these alternatives with the analogy of a ball tossed up, then slowing down during its ascent, whereas with equivalent velocity from a small asteroid, the ball may not ever stop.[87] Nave explains that over time, exaggerations must occur from even tiny variations from flatness and by now, great amplifications of any irregularities would have occurred.[88] Based on this model, if the Universe's density today is very close to critical density, that implies that in previous epochs the Universe had to have been much nearer to flat.[89]

There is evidence that geometrical curvature in the universe is almost flat. For this to accord with BBT's model of cosmic evolution since the Big Bang, an initial curvature would have to be restricted to an extremely narrow spectrum of possibilities, which seems highly unlikely.[90]

Monopole Problem

Grand unified theories are the only credible accounting for how in extremely high temperatures immediately following the Big Bang there was only one force. It had the features of electromagnetism, the weak and strong forces. This combined force inevitably resulted in very massive particles termed magnetic monopoles. Because these are electrically charged, they created a net magnetic charge within the individual particle. These should be one of the commonest particles today, yet no one has ever discovered one.[91] The fact that this is inexplicable by BBT is a massive problem for the theory, because it blatantly contradicts its most basic underlying principles.[92]

Debate on whether BBT Violates Conservation Law

In *The Inflationary Universe: The Quest for a New Theory of Cosmic*

Big Bang Theory Falsified

Origin, Guth uses the concept that gravitational energy is negative, and because it is in balance with the positive energy of matter, he concludes it is possible that the Universe evolved from "absolutely nothing" without violating any known conservation laws.[93]

This leads to several interesting questions. What is the mathematical or cosmological difference between "nothing" and "absolutely nothing." Guth is indicating that this is the case in his statement, so I am wondering how we would quantify and compare amounts of "nothing" from subatomic to greater than galactic quantities.

To argue that the BBT agrees with all conservation laws depends on accepting that energy exists as a positive substance in matter and a "negative" one in the form of gravitation. However, even if one were to accept that premise, it still does not account for how matter emerges from nothing.

Proponents of this pseudoscientific religious theory believe in a causeless genesis of the Cosmos in a "vacuum quantum fluctuation" between existence and non-existence prior to the Big Bang. Supposedly, the Universe emerged spontaneously from pure chaos. The idea of the Universe being chaotic arose because mathematics is incapable of modelling it in a logical way. The hypothesis of a vacuum fluctuation that preceded the Big Bang, first put forth by Dennis Sciama in 1969, eliminates the concept of the universe starting from a virtual particle appearing from an absolute vacuum and then transitioning to a real particle.

In 1973, Edward Tryon developed Sciama's hypothesis of a zero-energy universe emerging from a large-scale quantum fluctuation of vacuum energy. The result was a perfect balance between negative gravitational potential energy and positive mass-energy.[94] In the highly speculative theory, the pre-Big Bang "false vacuum," which contained energy, supposedly fell into a true vacuum state, thereby releasing the energy that powered inflation. A problem with this model is the idea of vacuum energy. Borchardt defines energy as being a calculated matter-motion term. I discuss this extensively in Chapter 8 under "Motion Required for Microcosms" Because energy refers to a formula involving mass, without matter there is no way to do this calculation. Thus, for the model to work there had to be mass, which requires the presence of matter in the false vacuum. Where did the matter come from?

There is considerable confusion over what the word energy

represents. Tryon claims that the universe adds up to a net zero. For his idea to work everything in the universe must add up to zero, which appears to be the situation in terms of all fundamental properties, such as electric charge and spin. Supposedly, all matter originated in the vacuum, created by the energy of gravitational potential energy.[95] Regardless of whether one supports the original Big Bang theory of the Universe
Not existing but having qualities of a particle that exploded in a Big Bang, or Tryon's quantum vacuum fluctuation, is of no significance in terms of conservation. The only difference is the first BBT cosmogony version blatantly violates the first law of thermodynamics, which states that the total energy of an isolated system remains constant, whereas the revised theory attempts to camouflage this violation. In fact, as explained earlier in this book, energy is not a thing, which means that this law needs to refer to the conservation of matter and its motion.

Tryon's quantum vacuum fluctuation involves the belief that a so-called false vacuum that contains unaccounted-for energy has the potential to become a real vacuum without any cause for such an event. The theory is merely an attempt to evade the very real problem of BBT's violation of conservation.

According to the zero-energy hypothesis, the Big Bang and inflation depend on a spontaneous vacuum fluctuation. QMT claims to show how this can occur. It claims that a vacuum has no matter or energy, but it contains "fluctuations". These are supposedly transitions between nothing and something, enabling a transformation to occur. Orthodox physicists objectify "energy" and imagine that with increases in energy, existing things appear out of that which is only a potential. QMT is based on the idea that particles are discrete "packets of energy" having wave-like properties. In this model, completely empty space has constant fluctuations of "virtual particles" and anti-particles, equally appearing and vanishing to balance one another. QMT experiments purporting to demonstrate this feature of vacuums is very misleading, because the aether permeating the mislabeled vacuums is not recognized. Based on Tryon's conception of "vacuum fluctuation," a void is not really the absence of everything, because it has the property of a potential. Understanding that only matter can have potential makes it clear that the void must be a form of matter.

The concept of a "void" as the non-existence of all forms of matter contains a contradiction. Stating that it exists, means that it must be something, which means it is material. It cannot be motion or energy, because it is not possible to separate motion from matter. The concept of "matterless motion" is absurd. It definitely is not pure "energy," because energy is just a calculation involving matter and motion. It is not possible for a thing to be non-material. The other option is to contend that there is no void. That implies that some form of matter, either baryonic or non-baryonic is present.

Most people likely think of a vacuum in the way defined by Merriam-Webster as "1. emptiness of space, 2. a space absolutely devoid of matter." However, QMT and relativity physicists use a completely different definition in which the absence of matter does not equate to emptiness because of the presence of hypothesized dark energy, neutrinos, gamma rays, gravitational waves, cosmic rays and virtual particles (a.k.a. vacuum fluctuations). Because all of these depend on the existence of some form of matter, including the mislabeled "virtual particles," there is no absence of matter in this description of the vacuum. Basic logic indicates that before the Big Bang, none of these could have existed without the presence of matter.

Based on Tryon's claim that matter was not present in the so-called false vacuum, there would be no theoretical possibility for motion or energy. Thus, to claim that the false vacuum "contained energy" is an illogical, absurd and impossible. Without the existence of a false vacuum, it could not fall into a true vacuum state, releasing its so-called "energy" to drive inflation. Therefore, the theory does not work.

So in complete contradiction of this definition of a pre-Big Bang perfect vacuum, it must have matter and therefore mass. This is because any property, including a potential for fluctuation, cannot exist in isolation to things, and there can be no reference to energy in the absence of matter and motion. This demonstrates that the true vacuum is only an imaginary construct.

Talking about the universe as "zero-energy" is subterfuge because it still involves a calculation. Where the energy calculation is possible, then matter must exist. Does that seem like a vacuum or void to you? Alternatively, is he claiming that these two "energies" did not appear until the Big Bang happened? If so then we still have the impossible situation of matter, with its measurable mass coming into existence out of nothing.

Referring to negative "gravitational energy" is a clever, but unsupported way of reaching an overall energy calculation of zero. However, the problem with this is that the apparent balancing of so-called positive and negative energy wrongly objectifies energy as a thing in "packets," which transforms between matter and energy. The concept is ridiculous.

To prevent BBT from violating conservation, its proponents argue that there never was any net energy created, because of the cancelling out effect of the negative and positive. Unfortunately, this idea derives from the confused concept that energy is a form of matter. In this incorrect concept, matter supposedly "contains" positive mass-energy, while gravity has negative potential energy. This idea emerged from not understanding that the energy concept is a matter-motion term for the calculating the exchange of matter's motion,[96] The formula $E=mc^2$, which involves a number for mass times the square of a number for velocity.is one example.[97]

The hypothesized "zero-energy" universe requires objectifying energy so that it inexplicably can transform into matter. This gives the pretense for the claim that there is a balancing of so-called positive and negative energy. However, if one realizes that energy is not able to transform into matter, then the zero-energy concept fails to solve the problem of the emergence of matter from nothing.

The assumption of "conservation" means that matter or its motion can never be the made or destroyed. This assumption has no known exceptions. There are no circumstances, under which it is allowable to violate this law and assumption of science. Thus, even the idea put forward by BBT supporters that the conservation law did not exist before the Big Bang is preposterous.

That suggestion does not have an answer to questions concerning when and how the law came about and what caused its appearance. Are laws a thing that exists apart from matter that directs its behavior, or are they just a description of the fundamental ways all microcosms behave, possibly as the result of being a property of matter? Because only matter can exist, what we refer to as laws do not exist in themselves, but instead represent an innate feature of matter, which enables us to predict matter's behavior. Thus, the argument that the creation of matter in the Big Bang did not violate conservation law is illogical, because matter and conservation law always exist together.

Also problematic for BBT is the fact that this theory needs zero-point energy. This is a property of the additional intergalactic space, which is supposedly being created constantly.[98]

Big Bang Theory Falsified

Bruce Nappi explains why BBT proponents do not appear concerned about any violation of conservation law with this comment: "Conservation laws do not invalidate Big Bang Theory because the mainstream theory was driven to be consistent with these conservation laws."[99] However, he goes on to state: The "challenge" needed for the BB was how to squash all the mass and energy into a singularity. That's their waving magic wand."[100]

Doppler effect and Redshift

The Doppler effect, also termed the Doppler shift, defines mathematical equations to explain an apparent change in wavelength observed for sound waves and electromagnetic radiation. This change in a wave's apparent wavelength and frequency is due to a relative motion of observer and source.

When a source is moving away from an observer, redshift occurs as the electromagnetic wavelength of radiation, including light, increases or shifts to the red end of the spectrum. These spectral changes in visible light's color from a galaxy are examples of the Doppler effect. Gravitational effects are also a contributing factor. BBT believers refer to "cosmological redshift," which they attribute to the stretching of space between galactic superclusters, as responsible for an expanding universe. They offer this as evidence for BBT. They envisage light as a wave-particle, an idea that violates the inseparability assumption for existing matter and its occurring motion.

According to the idea of a cosmological redshift, photons after a journey lasting almost 14 billion years are in pristine condition. The only effects noticed from the distance travelled are longer wavelengths. However, considering light as a wave within aether, supports an alternative explanation for redshift, as all wave motion is redshifted with distance. Thus, just as in any wave effect, the Doppler effect is a group effect.[101]

In his 1987 book, *Evolution of Quasars into Galaxies and its Implications for the Birth and Evolution of Matter*, Halton Arp referred to these objects as having large redshifts. He points out that once discovered in 1963; cosmologists immediately thought that they were in the outer regions of the observable universe due to interpreting their redshifts as resulting from moving at high velocities away from us. However, by 1966 there were indications that their redshifts arose from intrinsic factors caused by ejection from the centers of proximal galaxies. Without galactic redshift having a correlation to the Doppler effect, BBT loses it most important evidence, greatly impacting its viability.

Arp refers to further observations supporting this view, including those of young galaxies close to quasars. These galaxies also had high-redshifts without corresponding high velocity. He also references this data is in his book *Quasars, Redshifts and Controversies*, writing that the association of parent galaxies close to young, higher redshift objects has been confirmed.

Globular Clusters and Superclusters: Age Problem

Because BBT has recently been very definite in establishing a precise age for the Universe, the theory has an insurmountable problem reconciling its estimate with a cosmic age that is many times greater.

Globular clus*ters* are spherical groups of several million stars encompassing most galaxies, and orbit as a satellite around the galaxy's center. Cosmologists, seeking to understand the origin and evolution of galaxies, study these pervasive clusters. Based on BBT the Cosmos was still relatively young when the clusters developed.[102]

Astronomers deduce that Milky Way globular clusters to be up to 18 billion years old, which clearly contradict BBT's claim for the universe's age. Unless there is a serious flaw in stellar evolution theory, causing star ages to appear far greater than they are, then the universe is obviously much older than the BBT contends.

NASA's Spitzer space telescope discovered "old" stars and galaxies in which their light left at between 600 million and 1 billion years after the supposed Big Bang. This is far too soon after the hypothesized Big Bang for red giant stars in galaxies to have used up all of their hydrogen.[103]

Other observations discovered clusters and super clusters of galaxies at cosmological distances. These supposedly represent a time when the Cosmos was much too young to have given rise to such massive intergalactic structures.[104] There is no way for BBT proponents to explain these structures.

According to BBT, the age of the Cosmos is far less than the estimated 100 billion years needed for the universe to reach its present state.[105] Today's galactic superclusters are bigger than any structures that would exist as the result of gravitation based on the usual relative speed of galaxies [106] The same problem exists in explaining the great walls of galaxies, which are even vaster structures. Without relative velocities being considerably greater, there is no way that such large-scale structures could have developed in the given amount of time. BBT has no explanation for how this was possible.

In addition, making the problem more difficult for BBT is the fact that if velocities had been so much greater in the past, the tremendous energy involved in those would require some type of dissipation mechanism,[107] for which no evidence exists. The solution to the superclusters dilemma is to postulate a universe that is not finite, because there would be unlimited opportunity for the observed structures to form.

Age of Quasi-Stellar Objects (QSO) or Quasars

The requirement of several supernovae generations to build up a star's metal content forces the BBT to contend that the early Universe's stars, quasars and galaxies, which are called "primitives," were largely metal-free. However, this is contradicted by the observations showing even higher than solar metallicities (i.e., all elements other than hydrogen and helium) in the supposed earliest quasars and galaxies.[108] As redshifts get higher the iron to magnesium ratio increases. Astonishingly there is an absence of some line ratios, including iron abundance[109] between $z = 0$ and $z = 6.5$. Iron abundance at $z\sim6$ quasars is equivalent to its abundance in local quasars. In addition, there is far more dust in high-redshift galaxies and quasars than would be expected.[110]

In an unsuccessful attempt to accommodate these facts BBT proponents postulated that stars began forming earlier than previously thought, producing metals up to the solar abundance in about 500 million years. However, there is absolutely no supporting evidence for this imaginative hypothesis of rapid galactic evolution.

A paper in *The Astrophysical Journal*,[111] demonstrated that observations at high-redshifts are present in epochs when quasars are very young objects, and thus astronomers expect these to have low metallicity. However, the authors found that trends in the N V and O I broad emission lines (§ 3.2) suggest that metallicities are typically greater in higher redshift and more luminous QSOs. This contradicts BBT by suggesting the formation of a significant number of stars before or during the first appearance of quasars.

Quasars: Younger than Claimed

BBT contends that quasars were the most abundant during the earliest epochs,[112] but surveys refute this in revealing fewer quasars at both higher and lower redshifts. The number is up to 20 times fewer at z=5, which is equal to over 14.4 billion light years away.

Violates Second Law of Thermodynamics

This states that systems spontaneously evolve towards states of increased entropy. Contrary to this law, the Big Bang model proposes that rather than a condition of ever-greater disorder transpiring, the exact opposite occurred, which violates this fundamental physics law. If one assumes the big bang took place then the universe should consist entirely of evenly distributed particles. The fact that this is not the case is a falsification of BBT.

Uniform Absorption Lines

Astrophysicists refer to concentrations of neutral hydrogen in a quasar's spectra as damped Lyman-absorption systems. The hydrogen comprises the spectra from these systems that researchers detect at high-redshifts of 2 to 4 in quantity. These correspond to the period when galaxies were forming.[113] BBT has no explanation for these systems' absorption lines, nor does it account for why the relative abundances have so much uniformity.[114]

The Galaxy Formation Problem

A major unsolved BBT problem refers to two key points about the early expansion period. First, gravitation is not rapid enough to form galaxies and second, there were insufficient random non-uniformities to permit the gestation and development of galaxies. BBT theorists will have to be very creative to account for these problems.\

Galaxies Developed Too Early

In contradiction of the BBT prediction that no galaxies existed before the reionization epoch, a z=6.56 galaxy was discovered. Using the Abell 370 cluster, whose core holds several hundred galaxies 6 billion light years away, astronomers were able to magnify light from a galaxy behind this cluster, which is 15.5 billion light years from us.[115]

Using the detailed lensing model of this cluster, the researchers estimated a lensing amplification of 4.5 for this galaxy. The existence of this type of galaxy supports the view that the reionizing epoch is beyond $z=6.6$.[116] The fact that one galaxy of this type existed means many were present, because if only a few existed at the time, then hydrogen gas around them would have easily taken up their emissions, making them invisible.[117] Images revealed that two luminous radio-silent quasars are present at the centers of normal host galaxies. One is an elliptical galaxy containing. Quasar PHL 909 and the other is a spiral galaxy hosting Quasar PG 0052+251. BBT cannot explain why ionizing radiation is not disturbing either one of these galaxies.[118] BBT has no answer for the seeming contradiction of multiple quasars, which are highly redshifted objects, surrounding redshifted foreground galactic clusters. This quasar concentration is a mystery for Big Bang theorists, as they expect quasars to have a random distribution in the Cosmos. However, for those scientists who are not convinced that higher redshift indicates greater distance from us, the solution to the puzzle is that these objects are the same distance away. That was Dr. Arp's theory, which apparently was too contentious and threatening to BBT supporters when he proposed it in his 1972 book, *The Redshift Controv*ersy. In the introduction to the book, George Field refers to the fact that some astronomers doubt that expansion of the universe is causing all galactic redshifts. He emphasizes that if their view is correct then the Big Bang cosmology is questionable.[119]

CMB Temperature

Cosmologists estimate that 84.5 percent of all matter in the Cosmos is unknown dark matter. It derives its name because it does not interact with electromagnetism, which means that it neither absorbs nor reflects light.[120] There is conjecture that dark matter may contain supersymmetric particles. These would have partners with known particles. However, in an infinite universe, aether would be the non-baryonic matter that dominates. It heats up from radiation above 2.73°K, such as ultraviolet, X-rays, visible light and infrared light. Waves having temperatures less than 2.73°K and those greater than 2.73°K of the CMB could reduce galactic and intergalactic aether's temperature to the equilibrium temperature of the Cosmos. This would enable it to serve as an intermediate in dispersing radiation of that temperature, which could be the CMB. In this way, heat transfers from high-energy to low-energy radiation.[121]

Because intergalactic matter absorbs longer wavelengths more easily than other wavelengths, it indicates that microwave (mw) radiation, which is between infrared and radio, if not absorbed on the journey here, could not be uniform. Radio galaxies' ratio of radio to infrared intensities has to be due to some medium, such as aether between galaxies. This fact alone invalidates BBT because mw radiation cannot be originating beyond the galaxies. In addition, BBT cannot explain the variation in intensity in radio galaxies' wavelength. Before the CMB temperature was determined in 1963, estimations for it based on BBT predictions ranged from 5°K to Gamow's 50° K guess in 1961.[122]

BBT Stopped Predicting

There are no observations that provide validation for any quantitative prediction of the BBT. So-called successes have required continually increasing and changing its adjustable parameters after observations contradict the previous ones.[123] At least four other models—Plasma Cosmology, Variable Mass hypothesis, Meta Model Cosmology and Quasi-Steady State—have made accurate predictions.[124]

A valid, functioning theory makes testable predictions, but BBT advocates no longer even try to generate any. This enables them to avoid having to see the theory falsified, which in fact has already happened many times. To evade the admission that every failure of BBT is an invalidation of it, they simply formulated new modifications to add on to the theory.

A sound theory can have some refinements, but the basic theory remains intact and recognizable. This is not the situation with BBT. The profound changes it has undergone go far beyond small adjustments. Today's BBT with inflation, quantum transition of the vacuum, dark energy, a zero-energy universe and other appended abstractions is not similar to the theory that Lemaître proposed.

The necessity for these continual modifications and additions indicates that the theory is not fundamentally sound, which leads reasonable people to question it. This does not mean that a new cosmogony theory needs to replace it, as any such theory would also require violations of conservation. Only infinite-universe theories do not have this insurmountable problem.

Redshift Quantization

This is a theory that there is a tendency for quantized redshifts of cosmologically remote objects, such as galaxies, which means being at multiples of a specific value. Due to the relationship between redshift and distance, redshift quantization either can result from galactic distance from us or can indicate that the redshift is not a suitable way to determine cosmological distances. The significance of either of these being true would be devastating for the BBT.

Dr. Arp maintained that redshift is only secondarily due to an object's velocity and primarily caused by its relative youth. In Arp's understanding, two components determine the value of the observed redshift. There is an inherent one, determined by the object's matter, which changes over time. This is the largest factor in the total redshift of the quasar. The other component is the object's velocity.

Arp produced some observational evidence to support redshift quantization. Using large telescopes, he made images of multiple quasar pairs. Inferred from their having extremely high-redshift z values, they were supposedly receding very rapidly from us, which implies that they are very distant, ranging from 575 million to 13 billion light years. Arp's photographs revealed that they are physically associated with galaxies that astronomers know are relatively close to us with low-redshift values.

Arp's numerous photographs showed high-redshift quasars balanced on two sides of their low-redshift parent galaxies. These quasars are physically associated with low-red shift galaxies, ones that we know are close to us. The vast number of these pictures makes it very improbable that the occurrence results from random location or an illusion creating a seeming location that is incorrect.

The significance of Arp's photographs is enormous. It falsifies BBT by taking away the concept that cosmological redshift must be evidence of an expanding Cosmos resulting from the Big Bang event.

Contradictions in N-body Simulations

Astronomers use N-body simulations of a dynamic system of particles, usually under the domain of physical forces, such as gravity, to study processes such as filaments and halos of axes. These simulations with a possible non-zero cosmological constant and many variations of the BBT model do not correspond with measurement of optically selected galaxies. The surveys found two-point correlation function calibrations matched a nearly perfect power law over almost three magnitude orders of separation.[125] This contradicts the BBT.

Virgo's Streaming is Not Caused by an Attractor

In considering galaxies of a given brightness, having different redshifts on opposite sides of the sky, the unscientific Big Bang interprets these as evidence of a massive attractor pulling on them. Astronomers observe galaxies streaming out to 2,170 trillion kilometers on either side of us in a uniform direction in relation to CMB at distances over 423.8 million light years. It involves our local group of about 50 galaxies and the Virgo Supercluster of 40,000 members, which includes the Milky Way.

BBT cosmologists attribute this to the Shapley Supercluster, known as the "Giant Attractor," which includes more than 8000 galaxies having mass of ten million billion suns is racing toward Shapley, which is the most massive galactic cluster within a billion light years, at 2.16 million kilometers per hour. Using the discredited attraction theory of gravitation, cosmologists believe that Shapley exerts a pulling force on Virgo. This contradicts BBT, which has always contended that on these scales the universe exhibits uniformity and homogeneity. BBT misinterprets this redshift as representing widespread galaxy streaming. An alternative explanation that does not have contradictions is one that sees CMB moving in relation to us.

Microwave Radiation Fluctuations Data

Surveys from 2001 support Modified Newtonian Dynamics' (MOND) accurate alternative to BBT's dark matter explanation. It contends that differences between mass directly observed in the Cosmos and Newtonian dynamical mass may not be indicative of dark matter. Instead, inaccuracies in the limit of low accelerations in Newtonian dynamics may be causing it. MOND has correctly predicted spiral galaxies' rotation curves with a critical acceleration parameter. Additionally it has been able to adequately account for and explain systematic properties of spiral and elliptical galaxies.[126]

Big Bang Theory Falsified

Contradictory Densities Problem

Using the abundance of helium-4, lithium-7 and deuterium, BBT predictions for the total amount of baryonic (ordinary) matter provide very different results for each element and continue to worsen.[127] Critics of BBT refute its argument on the distribution of these light isotopes in the Cosmos.[128] In complete contradiction to BBT, when helium-4 (He-4), generated in massive stars by nuclear reactions, is added to the BBT prediction of helium, almost double the helium is produced than exists.

This led Lerner to conclude: "Thus either the blackbody spectrum or the light element predictions of the big bang are clearly wrong."[129] According to Lerner based on these wrong predictions, the likelihood that the Big Bang happened is smaller than one chance in a hundred trillion.[130] For any other theory that would be far more than enough to falsify it.

Entropic Cosmic View

Based on the Second Law of Thermodynamics (SLT), in closed, isolated systems entropy always increases. In a closed universe, this would result in it heading to heat death. However, observations of what astronomers refer to as the clumpiness–of-matter paradox, reveals a prevalent mixture of order and disorder. This violates SLT in closed models, thereby implying that the Universe is not closed. The apparent paradox vanishes in the absence of a finity assumption.[131]

While ignoring the obvious order at all observed scales, BBT attempts unsuccessfully to deal with the SLT violation by claiming that on a galactic-clusters scale, the Cosmos is homogenous, reflective of the expected disorder. However, if Alexander Kashlinsky and his NASA team's 2010 findings[132] have eventual confirmation, then these will have contradicted this BBT claim. They interpret their data to show dark flow of distant galaxy clusters inexplicably streaming out to a distance of 2.5 billion light years. Their findings differ from a group of scientists using data supplied by the Planck spacecraft of the European Space Agency who claim that the results of Kashlinky's team do not reveal any statistically significant evidence of dark flow.[133] However, Fernando Atrio-Barendela of the Planck team, disputes his team's findings, stating that we cannot rule out dark flow.[134] He declared that because his colleagues overestimated the measurements' uncertainty, the resulting analysis is flawed. Specifically, he maintains that it caused a subtle signal of possible dark flow to appear to be just noise. Going further, he suggested the data were consistent with the earlier findings from NASA's WMAP satellite that Kashlinsky's team used.[135]

Dark Flow Significance

Dark flow is significant because there is no observed matter in the Universe that accounts for it; indicating that a structure outside of the observable universe exists with gravitational effects on matter everywhere in the known universe. BBT never predicted this enormous mass beyond the observable universe, because it has no way to explain its presence. NASA astrophysicist Alexander Kashlinsky concludes that this mass indicates that the observable universe is a part of a multiverse that is very different from the known universe.[136] The Kashlinsky team's objective was to substantiate that with increasing distance, galaxies should appear slower, but instead they were surprised with their discovery that is changing our conception of the Cosmos and physics.

Kashlinsky reported his surprising results that the clusters were all moving at the same speed 3.2 million kilometers per hour toward a region between the Centaurus and Vela constellations. Astronomers do not consider this part of the hypothesized expansion of the Cosmos. Kashlinsky contended that although he had detected this dark flow only in galaxy clusters, it should apply to every structure in the known universe.

Interpretation of Dark Flow

Dark flow can best be interpreted by using observations of matter and motion patterns with the "Ten Assumptions of Science," and predictions of Neomechanics, which Puetz and Borchardt state:

"The local mega-vortex should appear as a massive spheroid extending well beyond the observable universe. It should contain a dense core at its center. It should possess a definite axis of rotation. It should contain materials with diminishing baryonic density, with distance form the core. And it should contain a high concentration of matter along its plane perpendicular to its axis of rotation. In short, a mega-vortex should behave as a massive spiral...We speculate that the dark flow indicates the spiral motion of the local mega-vortex."[137]

Big Bang Theory Falsified

Matter Concentration: Hundreds of Billions of Years

An enormous concentration of matter stretching an incredible 12 billion light years across, and requiring hundreds of billions of years to develop, appeared in surveys of galactic clusters in 2003.[138] This is a huge problem for BBT because its theorists are very definite that the universe is less than 14 billion years.

Completely contradicting and refuting BBT"s prediction that an earlier universe appears much different, discoveries revealed at the American Astronomical Society meeting in January 2004 showed that at high-redshifts the Cosmos appears similar. Thus, billions of years ago the Universe would have looked the same as it does now, which presents another huge problem for BBT.

BBT Needs Dark Matter

Dark matter is essential for BBT, otherwise indications for the universe's density are 20 times that which is calculated by the amount of light elements.[139] Physicists Luc Blanchet and Alexandre Le Tiec[140] suggest that gravitational dipoles of an unknown nature as dark matter can enable the inclusion of dark matter in MOND theory, which is a modification of gravity. They hypothesize that dark matter is a dipolar fluid composed of gravitational dipoles, which is a two-particle system. One particle has a negative charge while the other has a positive gravitational charge. Using a weak cauterization model of the mass distribution of dipole moments, they concluded that the dipolar dark matter reproduces MOND phenomenology at galactic scales. In this way, they suggested that dipolar fluid might be the non-baryonic dark matter.

Rejecting the existence of dark matter, Dragan Slavkov Hajdukovic[141] proposes that within the hypothesized quantum vacuum virtual gravitational dipoles produce a more powerful gravitational field. In an interview, Hajdukovic stated: "The key message of my paper is that dark matter may not exist and that phenomena attributed to dark matter may be explained by the gravitational polarization of the quantum vacuum."[142]

Hajdukovic's main concept involves a gravitational repulsion between anti-matter and matter due to particles and antiparticles having opposite gravitational charges. Therefore, within the quantum vacuum, virtual particle-antiparticle pairs are gravitational dipoles, and the vacuum is a dipolar fluid. The quantum vacuum's polarization by the gravitational field of the baryonic matter causes the illusion of dark matter.[143]

Orthodox physicists have denied the possibility of aether as a form of non-baryonic matter because relativity theory does not account for it. However, neither does dark matter, which has resulted in some researchers expressing the view that dark matter represents a failure of General Relativity, necessitating alternative ideas.[144] Once BBT proponents needed dark matter to save their theory, they accepted that non-baryonic dark matter is real without any direct evidence of its existence. I find it amazing that with all the speculation by orthodox physicists about whether dark matter is a mysterious form of non-baryonic matter, a dipolar fluid, molecular hydrogen or an illusion, they ignore the possibility of it being aether.

BBT Requires Inflation and Dark Energy

The Cosmos could not have reached a radius of 46.6 billion light years in the 13.799 billion years since the Big Bang. Because of massive discrepancies between what observations and what BBT claims, BBT theorists feel compelled to include "dark energy" and an "inflation field" even though neither is supported by any real evidence. To account for the diameter of the Cosmos BBT proponents concocted the fantasy of inflation.

Another problem appeared when the Hubble Space Telescope (HST) observations in 1998, using the questionable assumption that galactic redshift is entirely caused by the Doppler effect, indicated that billions of years ago the cosmic expansion rate was considerably slower than at present. This contradicted the BBT assumption that gravitation would slow the expansion.

Rather than admitting that BBT was fatally flawed, its supporters concocted dark energy. If you are wondering what dark energy is, you will have to remain in the dark, because even its inventors do not know.

At its website, NASA writes about dark energy: "More is unknown than is known. We know how much dark energy there is because we know how it affects the universe's expansion. Other than that, it is a complete mystery."[145] NASA then comments that it is an important mystery, because the universe is 68.3 percent dark energy.[146] NASA speculates that perhaps dark energy is a new theory of gravity, property of space, or a new dynamic fluid.[147]

A dark energy abstraction is crucial to BBT survival. Without it, BBT would indicate a universe age of about 8 billion years, which is 6 billion less than stated by BBT, and billions of years younger than that of many stars of the Milky Way. BBT requires an inflation field concept in order to conform to the observed CMB that is isotropic, which means it is smooth.[148]

Matter to Antimatter Ratio Problem

Based on BBT, in the early period of the Big Bang came the generation of all levels of matter, such as quarks and electrons, with their antiparticle forms in identical quantities.[149] Without any supporting evidence, BBT proponents speculate that most antimatter annihilated with matter due to some proposed asymmetry in the early universe.

This missing anti-matter cannot be in hypothetical antimatter galaxies, because their presence would have resulted in a detectable antimatter- matter boundary of visible gamma rays as Gary Taubes pointed out in a *Science* article in 1997 on cosmology.[150] He presented findings to show that anti-matter galaxies do not exist and ordinary matter dominates the universe. He reported on the results of the comprehensive analyses of three theorists who examined the pervasive gamma-ray glow and the physics of annihilation of matter-antimatter.

Astrophysicists and cosmologists have no definitive solution to why matter is dominant. BBT proponents theorize that after annihilation of paired particles and antiparticles following the Big Bang, there was some matter remaining, which was one-billionth of the paired particles. Astrophysicists make the unbelievable claim to have calculated how much matter remained after annihilation and that it equals the presently observable amount. It appears that these scientists calculated the former amount by assuming it must match the present quantity.

Cosmologist Rod Nave explains the problem with the above explanation of annihilation of antiparticles. He points out that following the commencement of the particle era, no process known to scientists can alter the net number of particles of the universe and after the first millisecond, and no change can occur in the balance between matter and antimatter.[151]

Too Many Faint Blue Galaxies

The number of faint blue galaxies at magnitude 28 is 10 times· higher than BBT predicts and indicates that spatial volume, rather than being smaller in earlier epochs was actually larger. Broadhurst, Ellis and Glaebrook refer to an "excess of blue galaxies" in their 1992 *Nature* article, *Faint galaxies: evolution and cosmological curvature.*[152]

Parameters for Elements Abundances

In reaching a determination for light-element abundances for each element, BBT requires one or more adjustable parameters, such as ascertaining a reason an element appeared or disappeared after the Big Bang. No real prediction was possible without these parameters.[153] Unlike BBT's insufficiencies in making predictions, an alternative model, known as plasma theory, accurately predicts abundance of light elements.

Quasar Redshift Brightness

In BBT, a redshift 0.1 quasar is about 10 times closer than a redshift 1 quasar. Due to the inverse law, a redshift 1 quasar, intrinsically similar to the other one, should be dimmer by a factor of 100. Using BBT's concept, that redshift equates with cosmological distance,[154] requires accounting for why it is equally bright on average. The strange explanation is given that quasar's intrinsic properties are evolving to become increasingly dimmer and smaller. This results in redshift 1 quasars being intrinsically 100 times brighter than 0.1 quasars. That is a very inadequate attempt to explain the observation of equal brightness of quasars.

By using an inverse-square law for the relation between a quasar's distance from us and its apparent magnitude, there is no need to resort to cosmological distance and expansion to account for the redshift. In general, large quasar redshifts do not correlate with distance, but are due to an intrinsic factor (primarily in most cases) mixed with a cosmological factor. In summary, improbably BBT requires quasars on average to become increasingly fainter to maintain an average apparent luminosity at all redshifts. This is a big problem for Big Bang theorists.[155]

Changing the Facts to fit BBT

In looking back to the Big Bang, if the ratio of critical density to actual density of matter has a difference from unity of one part in 10^{59}, then one can deduce that cosmic dissipation would have occurred long ago. When observations in the 1990s contradicted the BBT's view that expansion was slowing, and appeared to show the opposite was happening, its proponents were desperate for a way to keep some semblance of viability in the theory.

To enable BBT to be more reflective of cosmic observations, and have the facts conform to it, they appear to be following the idea of: "If the facts don't fit the theory, change the facts." Whoever made this comment, which is often misattributed to Einstein, clearly was being sarcastic, but BBT theorists frequently employ the idea. They considered that it was essential for the viability of their model to incorporate an added unknown energy which is said to be everywhere in space.

According to astrophysicists, dark energy is the "energy density of the vacuum of space"; a hypothetical "form" of energy infusing all space. (Note that these scientists are objectifying "energy.") Some of them think that this energy is the cosmological constant, which is equivalent to vacuum energy, or scalar fields. Energy is not matter, therefore there is no clustering of energy, but because physicists are confused as to what the word energy represents, they believe energy can dilute. They contend that this dilution occurs at a considerably slower rate than matter as expansion occurs. This supposedly enables it to act as a type of anti-gravity that causes the universe to expand.

The other variable introduced in a desperate attempt to preserve the BBT was the revived cosmological constant, considered by many physicists to be the simplest form of dark energy, as it is constant in time and space. This energy density value for the vacuum of space ceased to be real or necessary for Einstein in 1929 when astronomer Edwin Hubble's redshift data implied that galaxies beyond our local cluster were moving further apart due to expansion. In disrespect to the scientific method, these scientists function as if they can decide how much dark energy` exists, not understanding that energy cannot exist because it is just a calculation. They determine the amount based on what is required to maintain the Big Bang Theory.

Brightest Galaxy Clusters (BGCs)

This problem refers to the fact that BBT has no model for the evolving luminosity of the BGCs. Those having high X-ray luminosity are on average brighter at high-redshift,[156] while no evolution corresponds to those having low X-ray luminosity.

Constancy of Surface Brightness

BBT predicts that at cosmic distances, geometry is no longer applicable. As luminous sources get increasingly distant from us within non-expanding space, which includes the Milky Way, they appear smaller. Since redshift equates with distance, redshift and the angular size product stay constant. Also remaining constant is the surface brightness of objects, which astronomers calibrate as photons per second per unit sky area.[157]

Cosmology models that do not include expansion contradict BBT's prediction that there will be a decrease of $(z+1)^{-3}$ in surface brightness. Trying to account for this by claiming that galaxies were brighter in previous epochs is not the answer because it results in extrapolating ridiculously bright ones.[158]

HUDF and GALEX Data Conflicts with BBT

Hubble Ultra Deep Field (HUDF), and the Galaxy Evolution Explorer (GALEX) provided definitive conclusions on the test of surface brightness of galaxies to determine whether expansion is transpiring. Even if factoring in evolution, the possibility of an expanding cosmos is opposed to the data, which leads to the inevitable conclusion that the universe is not expanding. To a very high degree of certainty, surface brightness is independent of redshift.

BBT's Friedman-Robertson-Walker (FRW) expanding universe evolutionary theory needs high-redshift galaxies with impossible ultraviolet (FUV) surface brightness of a magnitude beyond the entire range of low-redshift galaxies.[159]

Quasars' Faraday Effect

On average, a quasar's greater distance should increase the Faraday effect (a.k.a. Faraday rotation), which is an interaction between a magnetic field and light in a medium. This occurs when quasars go through magnetized plasmas between galaxies, causing a rotation of the plane of polarization, which is linearly proportional to the part of the magnetic field in the propagating direction. In reality the mean Faraday rotation is greater near $z = 1$, where according to Arp they are at their intrinsically brightest, than they are at $z = 2$. The fact that redshift and rotation are not both increasing with distance is a problem for BBT theorists that they are not explaining.

In a 2006 paper, Peter Hansen[160] found that galaxies and quasars had redshift differences consistent with astronomical observations indicating apparent physical associations. Hansen showed it was possible to obtain high-redshift magnitudes for quasars in comparisons to the relatively low-redshift galaxies of their parent galaxies. Additionally he demonstrated that over a parametric threshold, only redshifts appeared in the quasars even though assuming isotropic ejection from galaxies.[161]

Since the publication of Arp's *Atlas of Peculiar Galaxies* in 1966, there has been a redshift controversy that BBT cosmologists have not been able to refute conclusively using statistics or through physical methods. The BBT cosmological hypothesis could not survive if even one of Arp's hundreds of examples proves to have a low-redshift galaxy with a physical connection to a quasar. Because of the enormous importance to BBT cosmologists of this not occurring, they contend that the observations must be in error, as they do not fit with their theory. Apparently, they subscribe to the idea mentioned earlier of changing the facts to conform to one's theory. This goes against the usual practice in science, which falsifies a theory when it does not match observations

Large Structures 1.65-trillion-years old

For the BBT to be viable, it is obviously not possible for anything to exist that predates the Big Bang. However, astronomers have discovered areas of space with diameters as large as 3.588,000,000 light years (1100 Mpc) that has very little luminous matter.[162] Based on the stated years since the Big Bang, the voids found in galaxy distortions could not exist unless galaxy velocities were considerably greater than they actually are. At the 600 km/s speed that galaxies typically travel, they could cover about 30 million light years (less than 9.2 Mpc). Dividing 1100 by 9.199333 and then multiplying by 13,799,000 equals 1.65 trillion years. Considering that 1.65 trillion years is about 119.6 times the BBT age of the Universe, the presence of these voids is more than enough evidence to invalidate BBT.

Cosmos is not Homogenous and Isotropic

BBT incorporates the cosmological principle. This principle derives from the idea that because scientists expect cosmic forces to act the same everywhere, there should not be large-scale structural irregularities in the matter originating from the Big Bang. Therefore, for BBT to be right, homogeneity must be present at great cosmological distances meaning that no structure can be any larger than 1.2 billion light years.[163] However, galactic surveys, instead of showing any signs of homogeneity, revealed galactic superclusters forming. In March 2016, astronomers discovered the BOSS Great Wall, a vast superstructure of over 830 galaxies that stretches as far as a billion pc across (1.3 billion light years or 11.7 billion trillion kilometers.[164]

As of July 31, 2017, the BOSS Great Wall remains the largest microcosm so far discovered. This enormous, vast supercluster of galaxies is possibly 10,000 times as massive as our Milky Way. Because most of the matter in the Great Wall may not be luminous, we may only be able to observe part of it. The existence of this enormous wall contradicts the homogeneity principle of BBT.

Some far away objects, such as gamma ray bursts and quasars, appear clustered together. These structures are so enormous that BBT has absolutely no explanation for them. BBT cosmogonists just claim that these objects do not have a linking mechanism so they do not belong together.[165] If an obvious, strong link ever becomes established, BBT proponents will claim it must be an illusion, because it conflicts with their model.

Hannes Alfvén stated: "I have never thought that you could obtain the extremely clumpy, heterogeneous universe we have today, strongly affected by plasma processes, from the smooth, homogeneous one of the Big Bang, dominated by gravitation."[166] I have not found a reasonable explanation from BBT proponents for this observation that contradicts their implausible theory.

Too Few Galaxies Change from Elliptical

If elliptical galaxies result from recently merged galaxies, then there should not be a distinction between angular velocities of stars at varying distances from the galactic core. A star's velocity would result in replacing an elliptical shape with a circular and smooth disk. BBT is unable to account for the absence of galaxies in the process of undergoing this transformation.

Big Bang Theory Falsified

Microwave Radiation Fluctuations

Surveys from 2001, support Modified Newtonian Dynamics (MOND) alternative concept to BBT's dark matter explanation. It contends that differences between mass directly observed in the Cosmos and Newtonian dynamical mass may not be indicative of dark matter. Instead, it may result from inaccuracies in the limit of low accelerations in Newtonian dynamics.

MOND has correctly predicted spiral galaxies' rotation curves with a critical acceleration parameter. Additionally it has been able to adequately account for and explain systematic properties of spiral and elliptical galaxies [167]

Maximum Energy Levels Exceeded

Due to the effects of CMB photons on objects, the absolute possible maximum energy is 60×10^{18} eV for anything going further than 20 million to 50 million parsecs. However, astrophysicists observe numerous particles having higher energies than that.

Temperature of the Medium between Galaxies

Scientists infer the intergalactic medium from small-scale falloffs of the power spectrum to be 20,000° K. However, evidence of evolution with redshift is missing. BBT's problem is that spatial expansion should bring a decrease in temperature, but this does not occur. BBT has no way to account for this. This decrease would not be from a transfer of heat or matter from intergalactic space that exists. This strongly suggests expansion is not occurring.[168] Without an expanding universe, there is no possible extrapolation back to a Big Bang event.

CMB Alignment with Local Supercluster

A conspicuous preferred orientation exists for CMB fluctuations, which for the octopole and quadruplet power are zero on a preferred axis.[169] Contradicting BBT's assumption of no preferred direction in space, the axis direction is along the Local Supercluster filament axis, which includes the Milky Way, and its direction corresponds to that of the Virgo cluster. This goes against BBT's assertion that CMB is isotropic overall, and that it originates very far from the local supercluster.

Impossible Mature Galaxies in Early Universe

In 2014 American, Australian and European astronomers discovered 15 improbably and amazingly mature galaxies at a distance of 12 billion light years. These were present when the Universe was supposedly less than 1.6 billion years old. For BBT to be valid requires that these galaxies were somehow able to produce an average of 100 billion stars, at a rate several hundred times faster than our galaxy does today.

In addition, the Milky Way is still making stars, but these galaxies had already stopped star formation. Caroline Straatman of Leiden University in the Netherlands stated: "However, in the distant past, galaxies were still actively growing by consuming gas and turning it into stars. This means that cosmologists expect mature galaxies to be almost non-existent when the Universe was still young."[170]

23 Billion-year-old Galaxies Contradict BBT

The Hubble Ultra-Deep Field (HUDF) uses Hubble Space Telescope's data to create a composite image of 10,000 galaxies in the Formax constellation. It reveals galaxies in a small area of space from between 400 million and 800 million years after the Big Bang, which is about 13 billion years ago.

In the following image of the Hubble Ultra Deep Field (HUDF) 2012 the data in this improved version of the HUDF reveals very distant galaxies, with redshifts between 9 and 12, as they appeared in the era from 350 million to 600 million years following the hypothesized Big Bang. The furthest object ever observed is in this image.

Big Bang Theory Falsified

Image 1. The Hubble Ultra Deep Field 2012[171]

Credit: NASA/ESA/S. Beckwith (STScI) and The HUDF Team.

Galactic structures as large as these would only form after 10 billion years.[172] Therefore, that means that these galaxies began forming at least 23 billion years ago. The 19 billion year old Milky Way globular clusters also contradict the BBT claimed age of the universe.

Astronomer Billie Westergard wrote about this situation in 2009.[173] He commented that according to BBT requirements the age of the cosmos must fall between 8.3 to 12.2 billion years. This calculation uses an expansion rate of between 55 and 80 kilometers/second over every million parsecs, which is 3.26 million light years. By 2011, this rate was narrowed to between 71.4 to 76.2 km/s/megaparsec.[174]

Increasing Variety of Specifications Needed

To sustain agreement with observations, models of the Big Bang require a continually increasing variety of specifications. For example, just in connection with conditions of origin and expansion, BBT needs a cosmic deceleration parameter. In BBT, this is a dimensionless measure of the cosmic acceleration of the expansion of space.

It also requires a cosmological constant, Λ, which is a pressure that resists gravity. Origin conditions also demand an expansion rate, referred to as the Hubble constant (h), and a density parameter, Ω, which is the ratio of matter density to the density required for a flat universe, divided into the density for baryonic matter and dark matter. Additionally, BBT proponents hypothesized dark matter (cold, hot and mixed) because their theory did not conform to observations. To make the BBT work, it needs to keep generating new features for the Universe.

CMB is Weak Evidence

BBT proponents have no good observational evidence for their theory that alternative, more credible theories can also account for in better ways. The most frequently used argument in support of BBT is the 2.7^0K of the CMB. This is supposedly due to a hypothesized photon-decoupling period about 380,000 years after the Big Bang. The theory proposes that when the Universe was 4,000°K, photons decoupled from matter resulting in infrared black-body radiation. The liberated photons moved freely without interacting with matter, and now we supposedly observe these photons as cosmic microwave background radiation.

Unfortunately, for the BBT, scientists can demonstrate that CMB can exist without concocting an absurd Big Bang. BBT supporters assume that the Cosmic Background Explorer (COBE) findings on CMB spectrum and isotropy resulted from the photon-decoupling scenario. However, in agreement with all observations, a non-homogenous model involving an intergalactic scattering and radio-absorbing medium[175] reproduces the same spectrum and isotropy of CMB.

Richard Lieu and Jonathan Mittaz,[176] physicists at the University of Alabama in Huntsville, were unable to locate gravitational lensing in the CMB. This is where one would expect to find it as it is supposedly the most distant light source in the Cosmos according to BBT. They concluded that isothermal sphere profiles and the NFW were not accurate descriptions of clusters, unless important elements of physics, responsible for shaping zero-curvature space, are absent in the standard cosmological model.[177] They state: "When all the effects are accrued, it is difficult to understand how WMAP could reveal no evidence whatsoever of lensing by groups and clusters."[178]

This was not the only problem their paper caused for BBT. It led to another one by proving Einstein's 1917 theory that the counteracting forces of gravity and expanding space at a "critical" density causes a "flat" universe. Based on his theory, if CMB originated from the Big Bang, it should not look as it does. Lieu thinks it is impossible for the radiation in CMB's cool areas to have journeyed nearly 14 billion light years to Earth with uniformity in sizes. Lieu states: "Einstein's theory of how gravity attracts light, coupled with the uneven distribution of matter in the near universe, says you should have a spread of sizes around the average, with some of these cool spots noticeably larger and others noticeably smaller, but this dispersion of sizes is not seen in the data. When we look at them, too many cool spots are the same size."[179] As Lieu explained the problem, too many cool spots are just right, with not enough expected size variation based on how an expanding Cosmos would distribute matter in the Cosmos. [180]

If the BBT accurately describes the Cosmos, then microwave radiation from some cool spots after transiting through mostly empty space, and dispersing because of the hypothesized "expanding universe," would appear small once reaching us. However, radiation from other cool spots that went near or around massive gravity lenses would seem bigger than average cool spots. For Lieu, the fact that no variation is present is a serious problem[181] because this indicates that there are no lensing effects occurring at all.[182] There are several possible explanations involving miscalculations in any of the following: for matter, the Hubble constant or in other cosmological parameters used in predicting the pre-lensed sizes of cool and hot spots in the CMB.

Another plausible one is that even though the radiation has come from far off, some cool-spot structures result from proximal physical processes, rather than being cosmological in origin. A third alternative is that processes near Earth produce the CMB, so the radiation coming to our telescopes never passes through gravitational lenses. This means it is not the afterglow of the Big Bang.

In research published in the *Astrophysical Journal Letters*, of April 10, 2005, Lieu and Mittaz use evidence from WMAP of a mildly "super critical" universe having more matter and gravity than the WMAP standard data interpretation indicates, which means serious problems for the BBT inflationary model.

The BBT argument that CMB constitutes evidence for the Big Bang is analogous to claiming that there is incontrovertible proof for a proposed theory of miraculous events (TOME). The evidence offered is the miracle of Santa Clause (SC). Believers in SC infer his existence from the presents that appear for children in hundreds of millions of homes every Christmas. Because SC's actions violate many basic laws of physics, he is an example of the TOME.

Although skeptics would offer alternative theories for the appearance of the gifts, such as that the children's parents are putting these under the tree, the TOME proponents would argue that there is no need for these unorthodox ideas, because SC is undeniable and we know that he brings presents to children every Christmas.

When people might ask TOME theorists how it is possible for SC to do everything he does, such as getting to so many homes in just a few hours and riding on a gravity-defying sled, the TOME supporters would have no explanation. They would simply state that these are inexplicable paradoxes, which are confirming proof of the TOME.

Even if TOME believers eventually had to give up the SC hypothesis, it would not shake their fundamental theory of miracles. They would simply come up with some other miraculous explanation for the appearance of the gifts. Only those scientists supporting TOME would get their research grants and have their papers published. Any scientist, including Nobel laureates, who attempt to dispute the TOME theory, would suffer very severe professional consequences. Using these methods, its proponents could rightfully state that there is a consensus of published scientists for TOME and marginalize anyone who disagreed with their absurd theory.

Conclusion: BBT is Invalid

From the evidence in this chapter, it is clear that the Big Bang Theory is invalid because it is fundamentally without any basis in observation. It violates several of the Ten Assumptions of Science, which for any other theory would make it non- scientific.

The primary argument of its supporters is CMB and redshift. This chapter provided several alternative explanations for both of these phenomena that do not require a Big Bang expansion model. Furthermore, the CMB origin theory and cosmological redshift are wholly dependent on two discredited concepts: an impossible inflation theory and the cosmological principle, whose supporters continually change their story when observations falsify it, making it non-falsifiable.

In considering all the issues with the Big Bang Theory, which requires so many parameters, and has multiple invalidating problems, if someone were just now proposing it, no competent scientists would take it seriously. There is a vastly superior, credible alternative cosmology theory without the many huge issues of BBT. It is Borchardt's 'Infinite Universe Theory'(IUT), which I describe and compare its characteristics to those of BBT in Chapter 11.

Research Funds Only Go to BBT Proponents

The development of alternatives to the BBT has had to deal with a great deal of intolerance for discussing them at mainstream cosmology conferences. Those who reject the BBT as invalid, because they realize that numerous types of observations have falsified it many times, feel compelled to be silent to avoid harming their careers.[183] This distorting filter is also used to determine whether observations should be considered accurate, based on the extent that they corroborate the BBT, enabling the anti-scientific dismissal of information on topics, such as the distribution of galaxies, redshifts, and abundances of helium and lithium[184]

Because BBT's advocates dominate the committees that control funding sources, almost all experimental resources and funding in cosmology goes to research on the Big Bang, which enables this theory to maintain its position in spite of facts that show it to be invalid.[185] Continual testing of a theory through observation is a basic and indispensable aspect of the scientific approach, but by only funding research that adheres to the BBT model results in a curtailment of this process and real research impossible.[186]

To deal with this problem, cosmology-funding agencies need to give consideration to supporting alternative models and to research on observations that contradict the BBT, while ensuring that those making funding decisions are not from the cosmology field.[187] These changes would facilitate scientific inquiry in developing the best theory that most closely conforms to observations.

Image 2. The magnificent starburst galaxy Messier 82[188]

Credit: NASA, ESA and the Hubble Heritage Team STScl/AURA M. Mountain (STScI) and P. Puxley (NSF)

References

---

[1] Borchardt quoting Gribbin in *The Scientific Worldview*, 194.
[2] Kaku, *Hyperspace,* 196.
[3] Ibid., 195.
[4] Ibid., 195.
[5] Ibid., 202.
[6] Odenwald and Fienberg,*Galaxy Redshifts Reconsidered.*
[7] Ibid.
[8] Ifl science, *What Is The Largest Known Object In The **Universe?***
[9] Science Focus, *What is the largest object in the Universe?*
[10] Borchardt, in private communication to me.
[11] Kane, *Are virtual particles really constantly popping in and out of existence? Or are they merely a mathematical bookkeeping device for quantum mechanics?*

[12] Ibid.
[13] Ibid.
[14] Odenwald and Fienberg, *Galaxy Redshifts Reconsidered.*
[15] Csep, T*he Universal law of Gravitation.*
[16] Zitzewitz , P., N., Robert F., Davids. M., *Physics Principles and Problems, 155*
[17] Merriam-Webster, *Definition of Inertia.*
[18] Borchardt, *The Scientific Worldview*,194.
[19] Alfvén, Quotations.
[20] People and Discoveries, *Big bang theory is introduced 1927.*
[21] Borchardt, *The Scientific Worldview*,195.
[22] Arp, *Additional members of the local group of galaxies and quantized redshifts within the two nearest groups.*
[23] Marmet, P., *Discovery of H_2, in Space Explains Dark Matter and Redshift.*
[24] Marmet, P., *Is Dark Matter Just Plain Hydrogen?*
[25] Ibid.
[26] Richter, Sembach , Wakker and Savage, *Molecular Hydrogen in High-Velocity Clouds.*
[27] Marmet, P., *Discovery of H_2, in Space Explains Dark Matter and Redshift.*
[28] Ibid.
[29] Marmet, P., *A New Non-Doppler Redshift.*
[30] Marmet, P., *Discovery of H_2, in Space Explains Dark Matter and Redshift.*
[31] Ibid.
[32] Marmet, Louis, *On the Interpretation of Red-Shifts :A Quantitative Comparison of Red-Shift Mechanisms* II.
[33] Borchardt, The Scientific Worldview, 97.
[34] Puetz and Borchardt, *Neomechanical Gravitational Theory Proceedings of the NPA (2012).*
[35] Borchardt, GPS Does Not Require Relativity.
[36] Puetz and Borchardt, *Neomechanical Gravitational Theory Proceedings of the NPA (2012).*
[37] Marmet. *Cosmic Microwave Background Radiation. Low Temperature Hydrogen in an Absolute Space as the Likely Cause of the Cosmic Microwave Background Radiation (thus no Big Bang).*

[38] Ibid.
[39] Guth, *The Inflationary Universe*, xiii, 12..
[40] Ibid, xiii.
[41] Greene, *The Fabric of the Cosmos: Space, Time, And The Texture of Reality.*
[42] Ibid, 286.
[43] Gribbin, *In the Beginning The Birth of the Living Universe*, 36.
[44] Ibid., 36.
[45] Greene, *The Fabric Of The Cosmos, 321.*
[46] Gribbin, *In the Beginning: The Birth of the Living Universe,* 37
[47] Hyperphysics, *Model of Earlier Events.*
[48] Greene, *The Fabric of The Cosmos: Space, Time, And The Texture of Reality*, 322.
[49] Ibid., 323.
[50] Ibid. 323.
[51] Ijjas, Steinhardt and Loeb. *Inflationary paradigm in trouble after Planck.*
[52] Iijas, Steinhardt Abraham and Loeb . *Inflationary schism.*
[53] LaViolette, *Is the universe really expanding?*
[54] Ibid.
[55] Djorgovski and Spinrad, *Toward the application of a metric size function in galactic evolution and cosmology.*
[56] Ibid.
[57] Hajivassiliou, *On the cosmological significance of the apparent of small interplanetary scintillation sources.*
[58] Nørgaard-Nielsen, Hans U.; Hansen, Leif; Jørgensen, Henning E. Salamanca, Alfonso Aragón; Ellis, Richard S; & Couch, Warrick J. The discovery of a type Ia supernova at a *redshift of 0.31.*
[59] Rigler, and Lill, *Infrared surface photometry of 3C 65: stellar evolution and the Tolman signal.*
[60] Stacy McGaugh. *A possible local counterpart to the excess population of faint blue galaxies.*
[61] Flandern, *Cosmology.*
[62] Sumner, *On the variation of vacuum permittivity in Friedmann universes.*
[63] Ibid.
[64] Lerner et al,. *UV surface brightness of galaxies from the*

Local Universe to z ~ 5.
[65] Lerner, Falomo, and Scarpa, *UV surface brightness of galaxies from the local Universe to z ~ 5.*
[66] Astronomy. *Universe is Not Expanding After All, Controversial Study Suggests.* quotes Eric Lerner.
[67] Ibid.
[68] Ibid.
[69] Ibid.
[70] La Violette, *Is_the_universe_really_expanding?*
[71] Ibid.
[72] Puetz and Borchardt, *Universal Cycle Theory*, 51.
[73] Davies, *Multiverse Cosmological Models.*
[74] Puetz and Borchardt, *Universal Cycle Theory*, 51.
[75] *Alfvén, Cosmology Myth or Science.*
[76] Ibid.
[77] Ibid.
[78] Ibid.
[79] Alfvén, *Cosmology.*
[80] Ibid.
[81] Ibid.
[82] Ibid.
[83] Ibid.
[84] Lerner, E. *Big Bang Neve Happened.*
[85] Hyperphysics, *Model of Earlier Events.*
[86] Nave, *Physical Keys to Cosmology: Difficulties with the Standard Cosmological Model.*
[87] Ibid.
[88] Ibid.
[89] Nave, *Physical Keys to Cosmology: Difficulties with the Standard Cosmological Model.*
[90] Csep, *Stars, Galaxies and Cosmology.*
[91] Ibid.
[92] Techie, *Top Ten Scientific Flaws In The Big Bang Theory.*
[93] Guth A., *The Inflationary Universe.*
[94] Tryon, *Is the Universe a Vacuum Fluctuation?*
[95] Cole, *The Hole In The Universe, 170.*
[96] Borchardt, Do Spent batteries have more mass?
[97] Puetz and Borchardt, *Universal Cycle Theory* 12.

[98] LaRocco and Rothstein, *The Big Bang: It Sure Was Big.*
[99] Nappi, B, in e-mail comment to me.
[100] Ibid.
[101] Borchardt, *Infinite Universe Theory,*
[102] West, Côté, Marzke and Jordán, *Reconstructing galaxy histories from globular clusters.*
[103] University of Alabama *New Look At Microwave Background May Cast Doubts On Big Bang Theory.*
[104] Ibid.
[105] Lerner, E., *The Big Bang Never Happened.* .
[106] Ibid.
[107] Van Flandern, Did *the Universe Have a Beginning?*
[108] Jain and Dev, *Age of High-redshift Objects—A Litmus Test for the Dark Energy Models.*
[109] Ibid.
[110] Ibid.
[111] Constantin and Shields, *Emission-Line Properties of z >4 Quasars.*
[112] Flandern, *The Top 30 Problems with the Big Bang Theory.*
[113] Wikipedia, *Damped Lyman-alpha system.*
[114] Flandern, *The Top 30 Problems with the Big Bang Theory.*
[115] Hu, Cowie, McMahon, Capak, Iwamuro, Kneib, Maihara, Motohara. *A Redshift z = 6.56 Galaxy Behind the Cluster Abell, 370.*
[116] Ibid.
[117] Ibid.
[118] Bahcall, Kirhakos, and Schneider, *The apparently normal galaxy hosts for two luminous quasars.*
[119] Field. Introduction *The Redshift Controversy.*
[120] CERN, *Dark Matter.*
[121] Gastra, *The Infinite Universe.*
[122] Flandern, *Cosmology: The Top 30 Problems with the Big bang Theory.*
[123] Rense.com *Big Bang Theory Busted By 33 Top Scientists.*
[124] Ibid.
[125] Gaztanaga and Juszkiewicz, *Gravity's Smoking Gun?*
[126] Sanders, *Modified Newtonian dynamics and its implications.*
[127] Lerner, *The Big Bang Never Happened.*
[128] Lerner, *Galactic Model of Element Formation.*
[129] Ibid.

[130] Lerner, E. *The Big Bang Never Happened.*
[131] Borchardt, *Resolution of STL-Order Paradox.*
[132] Kashlinsky, *Mysterious Cosmic 'Dark Flow' Tracked Deeper into Universe.*
[133] Atruio-Barendela, *On the Statistical Significance of the Bulk Flow Measured by the PLANCK Satellite.*
[134] Ibid.
[135] Ibid.
[136] Roach, *Unknown "Structures" Tugging at Universe, Study Says Something may be out there. Way out there.*
[137] Puetz and Borchardt, *Universal Cycle Theory*, 82.
[138] Lerner, E. *The Big Bang Never Happened.*
[139] Ibid.
[140] Hajdukovic, S., *Is dark matter an illusion created by the quantum vacuum?*
[141] Ibid.
[142] Zyga, *Dark matter may be an illusion caused by the quantum vacuum.*
[143] Hajdukovic, S., *Is dark matter an illusion created by the quantum vacuum?*
[144] Francis, *Do We Need to Rewrite General Relativity?*
[145] NASA, *Dark Energy, Dark Matter.*
[146] Ibid.
[147] Ibid.
[148] Rense.com *Big Bang Theory Busted By 33 Top Scientists.*
[149] Physics of the Universe, *The Big Bang and The Big Crunch.*
[150] Taubes, *Theorists Nix Distant Antimatter Galaxies.*
[151] Nave, *Physical Keys to Cosmology.*
[152] Broadhurst, S. Ellis, & Karl Glazebrook, Nature. *Faint galaxies –Evolution and cosmological curvature.*
[153] Lerner, *The Big Bang Never Happened.*
[154] Flandern, *Quasars: near vs. far.*
[155] Ibid.
[156] Flandern, *Cosmology: The Big Bang The Top 30 Problems with the Big bang Theory.*
[157] *The Big Bang Never Happened.*
[158] Ibid.
[159] Ibid.
[160] Hansen, *Redshift Components of Apparent Quasar-Galaxy*

Associations: A Parametric Model.
[161] Ibid.
[162] Jones, et al., *The 6dF Galaxy Survey: final redshift release (DR3) and southern large-scale structures.*
[163] IFL Science. *Newly Discovered "Cosmic Wall" Is 1.3 BILLION Light-Years Across*
[164] Ibid.
[165] Ibid.
[166] Hannes Alfvén, *Quotations.*
[167] *Benoit Famaey, Modified Newtonian Dynamics (MOND): Observational Phenomenology and Relativistic Extension.*
[168] Flandern, *Cosmology.*
[169] Lerner, The *Big Bang Never Happened.*
[170] Straatman, *Astronomy Magazine.*
[171] NASA, ESA, R. Ellis (Caltech), and the HUDF 2012 Team. *The Hubble Ultra Deep Field 2012.*
[172] Borchardt, elderly-galaxies-plague-big-bang-theory/
[173] Billie Westergard, *Dynamics of Black Holes and Structure Formation in the Hotson - Westergard Universe Model.*
[174] Discover, The Universe is expanding at 73.8 +/-2.4km/sec/megaparsec! So there.
[175] Lerner, *Intergalactic radio absorption and the cobe data.*
[176] Lieu and Mittaz, *On the Absence of Gravitational Lensing of the Cosmic Microwave Background.*
[177] Ibid.
[178] Ibid.
[179] Space Daily, *New Look At Microwave Background May Cast Doubts On Big Bang Theory.*
[180] Ibid.
[181] Ibid.
[182] Ibid.
[183] Rense.com *Big Bang Theory Busted By 33 Top Scientists.*
[184] Ibid.
[185] Ibid.
[186] Ibid.
[187] Ibid.
[188] NASA, ESA and the Hubble Heritage Team STScI/AURA M. Mountain (STScI) and P. Puxley (NSF).

Chapter 5

Absurd QMT, SRT and GRT

This chapter identifies and analyzes some of the numerous abstractions in QMT and relativity physics that do not correspond to the real universe and therefore have no appropriate role in any theory. These concepts include matterless motion, time is a dimension, energy exists as a matterless object, and spacetime. When critics of relativity theory question the validity of these incomprehensible, indeterministic abstractions, they confront significant challenge in trying to get physicists and cosmologist to realize that a problem exists. This is because the indoctrination into these concepts is pervasive throughout society, including all levels of the education system. Orthodox physicists, who realize that the abstraction of the spacetime continuum is impossible to conceive of, respond to pertinent questions about it by claiming that it is only a mathematical structure not a physical object. However, if mathematical structures are not representing any real thing or occurrence in reality, then they are entirely imaginary and therefore not reflective of the real universe. Spacetime represents an attempt to unify and objectify space and time, which is an impossible idea,[1] because time is motion not an object. "Space-time" in the Infinite Universe Theory is a deterministic matter-motion term that relates a thing's location to the motions of all things.[2] Space-time is a necessary valid abstraction that does not make the error of treating time (i.e. motion as a thing.

Copenhagen Interpretation (CI)

Niels Bohr and Werner Heisenberg developed this standard interpretation of QMT in the mid-1920s. Interpretation of quantum mechanics or the quantum interpretations of modern physics are other names for it. It originated as a mathematical formalism that had great experimental accuracy. In place of Newtonian conceptions, it proposes that reality involves fundamental discontinuity and probability.

Heisenberg was adamant about the observer's indispensable role in bringing the material world into existence.[3] He believed that even the quality of being does not properly belong to what QMT describes, but rather is simply a tendency or possibility for being.[4] In addition, Heisenberg stated: "In the Copenhagen interpretation of quantum mechanics, the objective reality has evaporated, and quantum mechanics does not represent particles, but rather, our knowledge, our observations, or our consciousness of particles."[5]

Heisenberg, commenting on George Berkeley's immaterialism or subjective idealism, indicates that matter requires detection to exist: "If actually all our knowledge is derived from perception, there is no meaning in the statement that the things really exist; because if the perception is given it cannot possibly make any difference whether the things exist or do not exist.[6]

He concludes that being perceived is equivalent to existence."[7] Immaterialism assumptions are clearly the foundation for this concept. The idea that the Cosmos had no existence prior to our observation of it strongly implies that the universe will vanish if we ever become extinct. I cannot imagine any more immature, implausible and delusional concept.

Those phenomena that are referred to as representing human consciousness are all forms of motion, which always require matter. It is impossible for matter to not generate life, and hence consciousness. Even non-animal species respond to their macrocosm, which can be considered a form of consciousness.

The standard model QM physicists believe that human consciousness is what enables matter to exist. That is nonsensical. It would be like saying that all motion would disappear if everyone stopped doing Tai Chi. That particular form of structured movement can end but other forms of human motion would continue, in addition to all the other motions in the Universe. If humans become extinct, other examples of life, which is a type of matter-in-motion, would still exist on Earth and throughout the Universe. It is impossible for matter to not generate life, and hence consciousness. To imagine that matter would cease if it did not produce experiences is a non-falsifiable, silly notion.

Carl Friedrich von Weizsäcker, a quantum physicist who was also a philosopher, reframed a basic CI idea in declaring that the Copenhagen interpretation means: "What is observed certainly exists; about what is not observed we are still free to make suitable assumptions. We use that freedom to avoid paradoxes."[8]

No scientific agreement, consensus or document on the meaning of the term Copenhagen interpretation exists. This fact was stated by John Cramer: "Despite an extensive literature which refers to, discusses, and criticizes the Copenhagen interpretation of quantum mechanics, nowhere does there seem to be any concise statement that defines the full Copenhagen interpretation."[9]

One area of agreement is the view that measurements influence systems resulting in a particular outcome, referred to as a wave function collapse. This links with the concept that before doing a measurement, physical systems lack definite properties and it is only possible to make predictions on probabilities that specific measurements produce specific outcomes.

Cramer identifies five concepts that are common to all versions of CI:[10]

1. Complementarity includes the wave-particle duality concept and making measurements causes specific things to take form through a wave function collapse.
2. Identification occurs between the state vector and "knowledge of the system/" The Schrödinger equation includes a series of state vectors, which change when observations occur, and over time represents the system's knowledge at various given times.
3. Statistical interpretation of the Schrödinger wave function as yielding the probability of particular outcomes in various states using the Born rule mathematical process.
4. Heisenberg's positivism involves an emphasis on considering only observable experimental results as opposed to the underlying "reality" or ascribed "meaning." This implicitly and at times explicitly, accepts the instrumentalism concept.
5. The Uncertainty principle means that there are limits to the accuracy level in simultaneous measurement of conjugate variables, such as momentum and position.

The concepts of CI involve absurd contradictions, which proponents try to evade by referring to them as "paradoxes." These are more than sufficient to invalidate it by any rational standard. In certain respects, it denies the reality of causality and among QMT proponents, there is no agreement on its complete status. The fact that there is any dispute on this basic scientific assumption, which any logical individual would accept as axiomatic, is shocking and demonstrates the ridiculous nature of QMT.

The idea of "non-causality" contradicts the fundamental basis of science, which is to find out what causes things and events. To reject a link between causes and effects is the antithesis of the scientific worldview. One cannot claim to be doing science while simultaneously denying causality. In seeing the absurdities in their non-causality concept, rather than accepting the fallacy of their interpretation of their observations that led to this, they blamed nature for the problem. Richard Feynman demonstrated this tendency in his quantum electrodynamics description of nature as absurd, which he claimed concurs completely with experiments.[11] He then asks us to accept the idea that nature is absurd.[12]

Heisenberg recounted how he had repeatedly asked himself in the early years of QMT development whether it was possible for nature to be as absurd as it appeared to be in the atomic experiments.[13] He concluded that the law of causality is not applicable to quantum theory.[14]

Encouragingly, some recent polls among physicists attending quantum mechanics conferences indicate that support for CI is declining. In 2013 polls, it received between 4 percent to 42 percent support as the preferred interpretation,[15] with the differences in results reflecting the particular approach of those invited to each conference. Nevertheless, it is clear that CI is no longer the dominant interpretation in quantum theory.

David Mermin thinks nature is magical: "while it is wrong to suggest that EPR correlations will replace sonar, it seems to me something is lost by ignoring them or shrugging them off. The EPR experiment is as close to magic as any physical phenomenon, I know of, and magic should be enjoyed."[16] Mermin adds that it is not clear whether we can learn any physics by pondering it.

A defense attorney using this sort of irrational thinking would not likely win any cases. Such statements would imply incompetence, resulting in having few, or no clients.

Space without Matter Abstraction

The relativity idea of space without matter is preposterous because it implies that non-existence is real. BBT proponents believe that the 2.7° Kelvin temperature of the CMB, which apparently permeates throughout the entire universe, is radiant heat remaining from a supposed recombination period of the universe, 380,000 years after the Big Bang. However, if the BBT is invalid, as the evidence in Chapter 4 indicates, then the temperature of CMB could be due to heat generated by the motion of aether,[17] or from molecular hydrogen. The estimated 100 to 200 billion, billion, billion stars in the observable universe[18] may also be contributing to it.

Impossible Motion without Matter abstraction

The idea that motion can occur independently of matter is another incredibly preposterous abstraction that does not have any association with any real thing, event, process or occurrence. During a science class in which I discussed this concept, a student of mine immediately appreciated its absurdity and laughingly asked me "What moves?"

Thinking that motion is separate from matter and treating it as such resulted from Newton utilizing an abstraction technique to devise a material point concept. After abstracting dimension and shape of any perceptible object, all that remains is just translational and inertial motion.

Understanding whether a word refers to motion or matter enables one to have clarity in the midst of great confusion about the meaning of these concepts. If the word is about a thing, then it refers to matter. If it is about what the thing is doing, then it is a motion term. Our understanding would improve by never using one of the words without including the other in the sentence. For example: "The tiger is running" is a correct use of the terms, but to say,: "there is running" does not have any meaning.

This abstraction of motion without matter is not logical or intuitive. The belief that it is possible comes from an assumption of separability; one envisions motion as existing independent of any specific thing. Rather than being properly understood as signifying what things do, in this concept one objectifies motion.

Abstraction of Energy

Energy is one of the most confusing abstractions used by orthodox physicists. As Glenn Borchardt frequently explains, energy neither exists nor occurs. This is a matter-motion term concerning the exchange of matter's motion representing a calculated result from a number for mass times the square of a velocity number. Energy is no more substantial than is the calculation for body mass index (BMI), which derives from a number for body weight in kilograms, divided by a number for height in meters squared.

The energy abstraction is misinterpreted by physicists as representing an actual thing that can change into matter and back again to an energy form. Accepting a wrong abstraction of energy involves an unwarranted presumption that energy is some type of immaterial, dimensionless thing. This idea is nonsensical because things always have boundaries, whereas energy does not.

Representative of the standard concept of energy is a definition in The Physics Hypertextbook: "The term energy refers [to] an abstract physical quantity that is not easily perceived by humans. It can exist in many forms simultaneously and only acquires meaning through calculation. A system possesses energy if it has the ability to do work. The energy of motion is called kinetic energy."[19] Nappi states; "Calculations do not confer meaning. Calculations give numeric precision to physics models. It is the models that produce the meaning. All the calculations do is provide precise and accurate measurements of interactions – if they use good models."[20]

Scientists also conceive of energy as independently occurring motion. The textbook *Physics Principles and Problems* reveals the difficulty orthodox physicists have with creating a reasonable definition for energy: "Non material property capable of causing a change in matter."[21] That is an absurd statement because there are no properties without matter, just as there is no motion without matter, as Borchardt has so often pointed out. Physicists may put forth the argument that fields have properties also. However, this does not refute my contention, because "a field is a physical quantity, represented by a number or tensor that has a value for each point in space and time."[22]

Merriam-Webster defines property as a: "a quality or trait belonging and especially peculiar to an individual or thing; *b*: an effect that an object has on another object or on the senses; and d: an attribute common to all members of a class."[23] Physicists may argue that second definition could easily be extended to the effect that fields have on other fields, or the effect of objects on fields etc. However, effects are always linked to objects, electromagnetic radiation (EMR) or fields, which are material quantities. EMR is defined by orthodox physicists as "energy that is transmitted at the speed of light through oscillating electric and magnetic fields."[24] However, I see no reason to think that EMR does not involve the motion of matter.

In classical mechanics, moving objects gain energy of motion, but in SRT, energy is equivalent to mass, and a moving body increases its mass through kinetic energy, which becomes measurable at high speed.[25] SRT contends that a body also acquires mass through the absorption of radiant energy. This concept erroneously objectifies energy. Stephen Puetz and Glenn Borchardt explain in *Universal Cycle Theory*, energy is a matter-motion term that refers to a calculation derived from using a number for mass, which is an outcome of the amount of matter involved, multiplied by velocity squared, which is based on the object's motion.[26] I agree with that definition.

Borchardt states: "Energy is neither matter nor the motion of matter. ...When ordinary matter is involved in the exchange of motion we use the matter-motion equation for kinetic energy $KE=1/2\ mv^2$. When EM (electromagnetic) radiation is involved we use the bi-directional matter-motion equation $E=mc^2$."[27] The fact that energy is a calculation provides the key to understanding the energy abstraction, although physicists seem to ignore this point while believing that energy is equivalent to matter.[28] This is a misconception.

It is important to realize that observable effects associated with the term "energy` result from the amount of mass associated with matter in motion interacting with other microcosms and its macrocosm. In understanding that "energy", simply refers to a mathematical formula, while realizing every formula is only a representation, it is clear that it is impossible for a formula to have any effect. The "energy" formula is simply a description of what transpires.

This is difficult for many people to understand due to having always thought of energy as a real thing. As an outcome of heavy conditioning about this concept in which energy is objectified, it will require considerable thinking for people to see that "energy" simply refers to a calculation. Because of the enormous confusion in defining energy, I agree with Borchardt's recommendation: "I suggest that the term energy be avoided whenever the more specific term matter or motion could be used instead."[29]

Petroleum as Energy Abstraction

Petroleum is a form of matter, and like all microcosms, it contains atoms, which are in motion. The potential of oil to transfer motion decreases in proportion to the internal motion of atoms transferring to microcosms performing work outside of the oil. This occurs with combustion. Using Borchardt's concept of energy as a matter-motion calculation, liquid petroleum provides the matter in the equation.

The motion of combustion transfers to engines enabling them to turn the wheels of vehicles. Although this motion is termed kinetic energy, I prefer to call it motion. The bond structure of atoms determines the energy calculation in a spring, whereas a completely different one is calculated in heat, electric and magnetic fields.[30]

I agree with orthodox physicists that "energy" is not a "thing" that has the ability to do work. But work is being done. In my view this is the result of microcosms-in-motion. Although motion can transfer between objects, the motion is always inseparable from matter, per the "fourth assumption of science."[31] Physicists measure the work done by motion in terms of joules.

Relativity physicists claim that matter can convert to energy as the formula $E=mc^2$ is misinterpreted to mean. Glenn Borchardt has written on the physical meaning of this formula. He explains it as the motion of matter converting from one level to another as an atom's inner, sub-microcosmic motion transfers to super microcosmic particles (termed aether) in the macrocosm, providing a medium for conveying that motion of matter as waves.[32] This explanation does not require the impossible abstraction of energy as a thing. As Borchardt explains in *The Physical Meaning of $E=mc^2$*, matter does not transform into energy, rather one type of matter in motion, changes into another kind of matter in motion and/or one type of matter's motion transforms into another kind of matter's motion. Additionally, it brought on the invention of both imaginary spacetime, and curved spacetime.

These abstractions are inherently contradictory. This is demonstrated in how spacetime refers to dimensions, yet these concepts without describing substances allow for curvature. The obvious question is "How can that possibly be?"

Inaccurate Interpretations of Energy and Force

It is incorrect to conceive of energy, force and momentum as being matter or motion, or of having existence or movement. These words refer to descriptions of matter in motion. They derive from multiplying a motion measurement, such as velocity, times one for matter, such as mass. Examples are these formulae: $E=mc^2$ for energy, $P=mv$ for momentum, and $F=ma$ for force. Energy, momentum and force are matter-motion concepts.[33]

All of these abstractions regarding motion, matter and energy depend on an assumption of separability. Using the definitions in orthodox physics, they do not represent real things or occurrences in any sense or at any level, so they are invalid abstractions that hinder the development of theoretical physics.

Definitions that objectify energy result in a ridiculous explanation for nuclear fission. QMT advocates and relativity physicists believe that an atom's inner motion converts into energy, which is weirdly conceived of as a matterless-object moving through the vacuum of space based on the totally unwarranted presumption that energy is an actual thing. This idea contradicts the definition of an object. Both the impossible matterless motion concept and energy as a thing abstraction are illogical and have no place in any scientific theory. Unfortunately, because physicists have been thinking in these terms for so many years, it will not be easy for them to abandon these concepts of how the universe functions. This is most regrettable.

Even physicists with magnificent analytical abilities fall into the trap of confusing abstractions with what is real. The greatest hindrance to scientific progress in physics is the plague of impossible, ridiculous and non-representational abstractions of reality. Orthodox physicists accept that these abstractions represent something that exists or happens, when in fact they are designating an imaginary thing and or occurrence. These proliferate in relativity physics, which in spite of their many profound paradoxes and contradictions are largely unquestioned as accurate descriptions of how the Universe functions. For example, consider the concept of space, which cosmologists envision as emptiness, rather than being a form of non-baryonic material in the form of aether.[34]

One cannot overemphasize that valid abstractions are attempts to act as a generalized category for things that exist or occur, but they do not have any reality outside of thought.

Wave-Particle Duality

In a preposterous interpretation of the double-slit experiment, QMT physicists conclude that light concurrently consists of a particle and a wave, having characteristics and behavior of both. This conclusion has its foundation in the idealism of CI. This abstraction requires a denial of realism, in which things exist independently of our observation. Being purely idealistic, it is therefore an anti- scientific interpretation.

Photons in a Vacuum

The idea of massless photons traveling through a vacuum of curved spacetime for billions of years at a constant speed with the only affect being a wavelength change is preposterous. Belief in this abstraction is limiting work on more rational models involving the propagation of light waves via the medium of aether.

Negative Mass

This section will present differing views on whether the idea of "negative mass" is valid.

Dr. Peter Engels of Washington State University and his colleagues[35] believe they observed negative mass on April 10, 2017. They surmise that it was created by reducing rubidium atoms' temperature to almost absolute zero, generating a Bose-Einstein condensate. They reversed the spin of some of the atoms in this state by employing a laser-trap. When the atoms were let out of the trap, they expanded and displayed what the researchers referred to as "negative mass," by accelerating towards a pushing force rather than away from it.[36]

Bruce Nappi agrees in the validity of the negative mass concept but accounts for it differently: "My SLT study came up with a very good model for negative mass. It has nothing to do with nuclear spin. The important factor in my "theoretical" discovery was observing that what science now calls "anti-matter" is really just anti-charge. In SLT, anti-matter truly has negative mass. It's behaviors are easily calculated by changing the sign of m in Newton's equations. But I find it hard for any measurements of single atoms in laser-traps to measure gravity changes because it is such a small force compared to all the other forces involved."[37]

> Glenn Borchardt explains this phenomenon using classical physics:
> "This interpretation that 'negative mass' has been produced is a wonderful illustration of regressive physics. It is true that the definition of mass is 'the resistance to acceleration.'... any impacts by supermicrocosms from the macrocosm will increase the mass of a microcosm... the momenta of the submicrocosms within the microcosm will increase. This is nothing more than Newton's Second Law of Motion acting across the microcosmic boundary. In other words, a "push" across that boundary will increase the motion of the insides of that microcosm before it accelerates the microcosm as a whole...the microcosm will tend to expand as those submicrocosms within the microcosm impact the microcosmic boundary on all sides. The regressive physicists involved in this study have disingenuously misinterpreted this push-back as "negative mass." It cannot possibly involve a decrease in mass, because any push-back by the submicrocosms within the microcosm only makes it even more difficult to accelerate the microcosm. Again, mass is the resistance to acceleration. Without acknowledging the submicrocosms producing this effect, regressive physicists are forced to invent yet another absurd ad hoc that makes no sense: 'negative mass.'"[38]

Gravity is not Distortion of Space

This is the crucial aspect of GRT. Einstein based this idea on the similarity between gravitation and all other types of acceleration, which led him to conclude that gravity bends light.[39] Einstein states: "With respect to the Galilean reference body K, such a ray of light is transmitted rectilinearly with the velocity c. It can easily be shown that the path of the same ray of light is no longer a straight line. From this we conclude that in general, rays of light are propagated lycurvilinear in gravitational fields."[40] The fact that astronomers observe light bending when passing close to the Sun does not mean that gravity, defined as a distortion of space, is the cause. This phenomenon near stars, which supposedly confirms GRT, has another explanation. Borchardt clarifies what is occurring: the light only bends when it is in the Sun's corona, which contains baryonic matter, just as the Earth's atmosphere does. Therefore, the light wave's behavior results from refraction,[41] which is an altered direction to wave propagation direction resulting from a change in the transmitting medium. This is supported by the fact that solar atmosphere density increases rapidly from very low values in interplanetary space to very high values nearer the Sun's surface.

Howard Hayden stated that "with a few lines of high school algebra" one can derive refracted starlight," This accuracy level is exact; whereas the tensor calculus and Riemannian geometry of general relativity produce only gives only an approximation.[42]

Gravity, Not Mass Increases with Velocity

Because the term "relativistic mass" has misunderstandings and controversies, Peter E.Manor[43] proposes an alternative concept. He contends that the greater force needed to accelerate objects traveling at increased velocity is an outcome of greater inertia. Thus, it results in more gravitational force required to "pull" that object, which leads to an increased gravitational constant. Manor derives a formula to calculate such variations in the gravitational constant, thereby eliminating the necessity of the term "relativistic."

Manor concludes that speed and mas determine an object's inertia. It becomes increasingly more difficult to accelerate, change an object's course or decelerate with greater "energy." Relativists account for this phenomenon as caused by the object's increased relativistic mass, which unlike the unchanging invariant mass ("rest mass"), it increases with velocity. The problem with this explanation is it requires change in the object's properties. So Manor proposes a different explanation for why an object's velocity brings increased resistance: it relates to changes in gravitational force, and thus in the gravitational constant.[44]

Manor states that the gravitational "pull" has to be at least of the same magnitude as that of the inertial force to overcome it. If an object with constant mass travels at an unchanging speed, the Earth's gravitational influence is always related to the object's velocity as is stated in the equivalence principle regarding inertial mass and gravitation. They have equal force on an invariant mass moving at the same speed. The gravitational constant depends on the pace of the mass. It only appears that object's mass changes when its inertia alters.[45]

Scientists who Dissent with Relativity

In a dissenting view from relativity, published in Infinite Energy magazine, William H. Cantrell, Ph.D. of the University of Texas contends that nuclear fission does not depend on relativity. He points out that Einstein borrows from Lorentz and Poincaré's math to modify time and length so that there is apparent constancy in the speed of light to for all observers. Cantrell illustrates the circularity of Einstein's logic and the predicament of falsifying it with an example of a completely transparent hypothetical second Earth moon composed of green cheese. It would be impossible to disprove this theory by trying to illuminate it.

Petr Beckmann, Ph.D., founder of the Galilean Electrodynamics journal (now edited by Cynthia Kolb), wrote textbooks on electrical engineering and taught at the University of Colorado. He reinterpreted relativity in his 1987 book *Einstein Plus Two*. He argues that the observations supporting relativity can be reinterpreted in a much simpler way that maintains universal time. The journal publishes articles that dispute relativity, especially those showing that it leads to logical contradiction, is unnecessarily complicated, and has confirmation in only in a narrow sector of physics. One of those papers was by Ruggero Santilli founder of the Institute of Basic Research, who highlighted nine inconsistencies between GRT and current scientific knowledge, such as quantum electrodynamics, which is one of the most experimentally verified theories.[46]

Tom Van Flandern, Ph.D. was a special consultant to the Global Positioning System (GPS) that use atomic clocks for ground observers to ascertain a satellite's location. Prior to GPS's launch SR theorists were unsure it would function. Flandern posed several questions that relativity proponents have not been able to adequately answer: "Why do photons from the sun travel in directions that are not parallel to the direction of Earth's gravitational acceleration toward the sun? Why do total eclipses of the sun by the moon reach maximum eclipse about 40 seconds before the sun and moon's gravitational forces align? How do binary pulsars anticipate each other's future position, velocity, and acceleration faster than the light time between them would allow?

In a paper in Physics Letters, Van Flandern claims that gravitation propagates 20 billion times faster than light, contradicting SRT.[47] I mention it despite rejecting that gravity is something that travels through space, because even that speed is not sufficient to explain observations.

Van Flandern states that physicists are taught that the speed of light in a vacuum cannot be exceeded, but are also trained to conceive of gravitation as instantaneous. As a result of this contradiction he rejects the present interpretation of relativity theory. Having the speed of light as the limit, then gravitational influences would be delayed if one assumes they propagate through space. This is required because if it moves at the speed of light Earth will travel for 8.4 minutes by the time the Sun's theorized "attractive" force reaches us. Because this causes Earth's and Sun's "pull" to not be in the same line, these misaligned forces would "double the Earth's distance from the Sun in 1200 years."[48] Since planetary orbits are stable, Newton held the immediacy of gravity's effects as being an axiom.

Astronomical studies support this conclusion. Earth accelerates toward the actual instantaneous direction of the Sun, which results in sunlight and gravitational effects coming at us from slightly different direction. This indicates different light and gravity having different transmittal speeds.

Observations that appear to support SRT or GRT can be more simply explained by assuming the presence of aether Alternative theories to relativity generally involve this inclusion—enabling the transmittal of electromagnetic waves—hypothesized to correspond to a "gravitational field" surrounding all celestial bodies. Most of these theories propose the aether gets denser with greater proxy surface of bodies. However, Glenn Borchardt and Stephen Puetz describe the situation in reverse to this in their Neomechnaical Gravitational theory, which is similar to Newton's gravitational pressure gradient.

GRT was supposedly confirmed by its success in predicting the advance of Mercury's perihelion, which is the orbit point nearest to a star. Paul Gerber had published the identical equation 17 years previously, which was before relativity's creation. In not acknowledging Gerber's work, even though it had been in a book Einstein had studied, he lied by saying he had never seen it.

In studying the calculations, Flandern discovered "three separate contributions to the perihelion; two of which add, and one of which cancels part of the other two; and you wind up with just the right multiplier."[49] When Flandern asked one of Einstein's former colleagues from Princeton's Institute for Advanced Study, how he thought Einstein derived the right multiplier, he was told: "knowing the answer," Einstein "jiggered the arguments until they came out with the right value."[50]

For GRT to be valid, it must have universality. However, Flandern shows this to be contradicted by the orbits of binary stars and their unequal masses, which relativity does not account for. Van Flandern writes that the only defense from physicists to this problem is that this indicates "something peculiar about these stars, such as an oblateness, or tidal effects." The more probable reason the formula is not applicable everywhere is that in his *Explanation of the Perihelion Motion of Mercury from General Relativity Theory* Einstein chose independent variables that would produce the result he required to "explain" Mercury's orbit. Einstein's footnotes contain a statement about the change of the theoretical premises. He assumes certain conditions are generally true and then states he is entitled to add to his previous field equation another limiting condition.[51]

Van Flandern posits that everything is understandable by assuming the pervasiveness of "a light-carrying medium. ... it takes longer for each electron in the atomic clock to complete its orbit."[52] Thus fewer "ticks" are made over the same period than occur in a stationary clock. Time does not slow down but moving clocks are slower due to ploughing through an aether medium. (Actually, it is impossible for time to "slow down" because being motion rather than a thing, it cannot move.) Every experiment that apparently confirms SRT is done on the Earth's surface, where each moving particle or moving atomic clock is slowed from pushing through Earth's gravitational field. Because the gravitational field is thinner 20,000 kilometers above the Earth, clocks operate 46,000 nanoseconds per day faster than those on ground level. Orbiting clocks also move through that field at their orbital speed, which is three kilometers per second, causing them to function 7,000 nanoseconds slower each day than non-moving clocks.

GPS designers compensate for this by reducing clock speed by 39,000 nanoseconds a day prior to launches. This enables these orbiting clocks to operate at the same rate as ones on the ground and thus the system works as required, allowing ground based scientists to locate satellite position precisely. Because clocks in orbit move rapidly, and relative to observers on ground have different speeds—which according to SRT the speed relative to observers is the relevant one—scientists anticipated making constant relativistic clock rate corrections. However, despite not making these adjustments after takeoff, the system functions perfectly, providing a contradiction to SRT. Since relativistic expectations do not agree with current data relativity supporters are claiming that to make

SRT work only requires another "reference frame."

This problem does not exist in the alternative theory known as modern mechanics developed by Steven Bryant. As described in Chapter 6, it accounts for all the facts with a higher predictive accuracy level than relativity theory. With the vacancy created by the falsification of SRT and GRT, I believe it is an excellent candidate to be the replacement.

Herbert Dingle of the University of London demonstrated SRT's illogic with the example of two observers with clocks: when one moves relative to the other the one moving runs slower than the other one. But the principle of relativity claims that when one thing moves in a straight line from another, either one can be seen as moving. Thus, if clock A is moved, it runs slower than B, and clock B runs slower than A.

SRT tries solving this problem using Lorentz transformations math. This uses inertial frames, which are a set of measuring devices moving with constant velocity along a straight line and without rotation. It results in the idea that a moving clock runs slow. "But …it is a condition of the problem that either clock can be regarded as the 'moving' one, so this … solution …requires equally that A works faster than B and that B works faster than A, … we know enough about clocks to know that one cannot work steadily both faster and slower than another."[53]

Alan Lightman gives a non-solution in his 1992 edition of *Great Ideas in Physic,* which is omitted in the 2000 edition: "The fact that each observer sees the other clock ticking more slowly than his own clock does not lead to a contradiction. A contradiction could arise only if the two clocks could be put back together side by side at two different times." However, he points out the problem is avoided because clocks in constant relative motion in a straight line "can be brought together only once, at the moment they pass."[54] Although the theoretical impossibility of testing this theory defends it from its inherent contradiction, it also disqualifies it from being considered scientific.

Physical Lorentz Relativity versus Observer-Centric SRT

Hendrik Lorentz formulated his Relativity theory before Einstein's SRT. Although his equations *appear* the same as those of SRT, their physical meaning differs by using the concept of absolute velocity rather than SRT's relative velocity. Lorentz provides one coherent physical description of space and time. In his relativity, a clock's rate slows ("clock retardation") with increased velocity, which corresponds to what is occurring in the world. Because all observers have an identical view of what is physically taking place, there are no contradictions or so-called "paradoxes" to contend with. Contrasting with this, SRT has symmetric equations, which means "all inertial frames are equivalent,"[55] and produces a vastly different physical meaning. Because SRT is observer-centric, where each one has a unique view of spacetime, which conflicts with all other views, it does not describe what is physically occurring.[56] For example it maintains that that clock A can run slower than Clock B, which ticks slower than Clock A. SRT's symmetric equations are a function of relative velocity and describe an effect that is "just observed," in reference to physical quantities such as mass or length, or in the idea of accumulating "time," Within specific domains, such as particle collisions at relativistic speeds, SRT has accurate numerical predictions.

How SRT Derives the "Correct Result"

Although SRT's apparent descriptions of what transpires vary to a large extent, and are fairly poor physical representations, for particle collisions and much of kinematics, its predictions of physics results are accurate. This is because its predicted outcomes are constrained by other factors to be compatible with what physically occurs in broad areas.[57] In Special Relativity, each inertial user views the world as if in the preferred frame using Lorentz Relativity. Despite this view being generally incorrect, conservation law constraints guarantee that the predicted result will be as accurate as the Lorentz final result predicted by Lorentz Relativity.[58] An example occurs in an accelerator collision of two particles, where a third particle may result from the kinetic energy. Even though the SRT descriptions have large differences, suggesting that two SRT observers would disagree on whether a third particle exists, conservation laws require that "available" kinetic energy equates to the quantity of "available" absolute kinetic energy in Lorentz Relativity.

SRT's implied physics descriptions are inaccurate despite its ability to predict final outcomes within its domain. SRT often implies that specific effects, which are a function of absolute velocity, are a function of relative velocity. Actual physics involves absolute kinetic energy and absolute momentum, so SRT abstractions of relative kinetic energy and relative momentum are mere bookkeeping ideas.[59]

Because every inertial SRT observer maintains that their yard stick is the longest and their clock is the fastest, they all seem to be in Lorentz

Relativity's preferred frame, which directly describes what's occurring physically, providing we know what the local preferred frame is. Even though SRT's math is misleading regarding specifics of what's physically transpiring as a result of simulating Lorentz Relativity, which is the right physical, and due to conservation law constraints and Lorentz factor math properties, SRT can predict results of certain events.[60]

A problem occurs when physicists use SRT as if it directly describes "what's happening physically." This thinking mistake results because SRT appears to accurately predict physics outcomes and the scientists are not familiar with any theory that directly describes "what's happening physically."[61]

SRT's Problem

In excess of 3,500 papers and books catalog many problems for SRT including inconsistencies with empirical data and contradictions. Percival explains that SRT's problems include not being clearly defined, being interpreted in several conflicting and mutually exclusive ways, and having its practitioners use it outside its poorly defined valid domain [62]

Another problem is that orthodox physicists fail to notice that there is a large area of physics where what SRT would predict is totally incorrect. For example, SRT's concept that "all inertial frames are equivalent," dissolves in the "time accumulation domain. This is because the actual physics of proper time manifests as an absolute accumulation and proper time is observer independent, absolute and operationally defined.[63]

GPS reveals what the preferred frame is (locally) and proper time accumulations provide data from A's viewpoint and B's viewpoint. The data demonstrates an absence of symmetry for what's happening physically, for proper time accumulation physics, and for what's observed by A & B. GPS data proves that the SRT model specifically cannot be used, and is not used in GPS, because it is not accurate in describing what's happening physically.[64]

Because the equation used in GPS appears similar to SRT's time dilation equation, relativists often claim that GPS depends on relativity theory, which is not true as the actual physics of GPS's clocks is not in SRT's domain. SRT is constructed using relative velocity, kinetic energy, relativistic mass and observed time of clocks. However, to have a proper theory of what's going on physically, including proper time accumulation, one needs the constructs of absolute velocity, absolute kinetic energy, absolute rest mass and proper time for all clocks.[65]

In contradiction of SRT, GPS uses the Lorentz physical model for velocity dependent proper time. Because the two theories use equations in which the mathematics *appears* identical, physicists mistakenly think they are the same. Although each theory is based on the Lorentz Transformation (LTs) equations, their interpretation of those is radically dissimilar.[66] In Lorentz Relativity, LTs are only used in transforming to a non-preferred frame from the unique, preferred one. It also uses LTs <u>true inverse</u> to transform from non-preferred frames to unique, preferred framed. Thus, the use of the LTs describe <u>physical</u> differences between the unique, preferred frame and the non-preferred frames and are not symmetric. The "v" represents absolute velocity as determined in the unique, preferred frame.

SRT physicists assume equivalency of all inertial frames and use LTs symmetrically to transform between any two of them. Because they use symmetric relative velocity, whatever clock A determines about clock B's speed, B will find the same for clock A. Every inertial observer falsely considers themself as occupying the unique, preferred frame.

Open Letter to the Physics Community: The Twin Paradox[67]

2011 is the centennial anniversary of the publication of Paul Langevin's famous paper "On Space and Time" in which he introduced, what became popularly known, as the Twin Paradox. This letter discusses the results of a recent study of the Twin Paradox problem. This study concluded that, after 100 years of work on this famous problem in special relativity, the Twin Paradox continues to be unresolved. Our purpose in writing this letter is to request that a specific, new course of action be undertaken to resolve this problem.

Follow-On Open Letter - On Special Relativity

The responses to this Twin Paradox Open Letter showed that there was indeed great confusion about the meaning of time dilation and other key constructs in Special Relativity even for the official spokesmen for Special Relativity and for the most prestigous of mainstream physicists. So we have created a follow-on. This follow-on open letter simply asks for the physics establishment to clearly define the meaning of the key constructs of Special Relativity. If Special Relativity is well understood, this will be a trivially easy task to do. On the other hand, if the mainstream is unwilling or unable to clearly define the meaning of Special Relativity, that strongly reinforces the argument for such a definition.

Intermediate Results, Summary, Conclusion

The open letter was posted in October 2010. In addition, we sent the survey of Twin Paradox questions to physics organizations, physics journals, university physics departments and individual physicists. A summary of results follows:

- A significant number of the physics community, including the signatories below, believe that the Twin Paradox is unresolved or unresolvable with currently accepted theory. Support for the "There is no paradox" position comes primarily from the "core of the mainstream", namely, those who control what will be published, funding, etc.
- There's virtually no support for Special Relativity's time dilation as cause for the net proper time dilation (NPTD) and, according to survey results no such support from the "core of the mainstream". (Overlooked is the fact that the exact same logic that rules out Special Relativity's time dilation as cause for the NPTD also rules out interpreting Special Relativity's time dilation as describing any related physical effect (e.g., any NPTD).) This retreat from Einstein's original 1905 claim by virtually all, including Einstein himself, is due to that claim leading to obvious contradictions. Despite this dramatic reversal, support for the "There is no paradox" position remained as confidently held as before. Proponents of the "There is no paradox" position went on to endorse a wide variety of (mostly mutually exclusive) alternatives over the decades
- The "core of the mainstream" position has now evolved to the very general claim that the NPTD is due to the "kinematics of spacetime". Further, this very general claim includes the assertion that one cannot tell how much of the NPTD accumulates between events unless both of a pair of events lie on both twins' worldlines (e.g., just the start and end events of a Twin Paradox round trip). It does indeed appear that currently accepted theory, namely relativity, cannot describe the physics (e.g., the

how, when, where) of the NPTD accumulation. Further, the claim that the final, total NPTD is due to the "kinematics of spacetime" does not include any specific characteristic that would tie it to either Special Relativity or General Relativity. In fact, logic indicates that describing how the NPTD accumulates requires the construct of velocity with respect to a single physics frame or object or field.
- A preferred frame theory, which employs the same Lorentz factor as Special Relativity but where velocity is always measured with respect to a single, preferred frame, gives a complete physics explanation of how the NPTD accumulates without any hint of a paradox
- It's claimed that existing experimental data regarding time dilation supports relativity. This appears to be true for effects due to a difference in gravitational potential. However, Special Relativistic (velocity related) data falls into one of the following categories:
a) The data is indeed consistent with Special Relativity but is also equally supportive of preferred frame theory.
b) The data approximately matches Special Relativity but is a better match with preferred frame theory.
c) The data appears to be consistent with Special Relativity but ONLY if velocity is measured with respect to a specific preferred frame.
d) The data is in sharp disagreement with Special Relativity but agrees with preferred frame theory.

An example of the above is GPS. For determining clock rates as a function of velocity, special relativity is NOT actually used. The model and algorithm actually used is from Lorentz Relativity where all measurements and calculations use velocity with respect to a single frame - in the case of GPS, it's the ECI frame. Further, while Earth observers measure, in effect (see Report for the full details), satellite clocks as running slow due to velocity, clocks/instruments on the satellites measure earth clocks running FAST due to velocity. One might try to rule out these latter observations that are in sharp contradiction with special relativity's time dilation by claiming such observations are invalid as the satellites are not inertial, but this assertion would also rule out using special relativity for the GPS system for the velocity component as neither satellites nor the rotating earth are inertial. Another example of the a) class of data discussed above is kinematics including particle collisions. For this area, as many have noted, the

predictions from Special Relativity and Lorentz Relativity are the same. So the Einsteinian relativist might ask, "*Why irritate me by bringing up preferred frame theory and Lorentz Relativity?*" The reasons are:
1) Special Relativity does not directly describe what's happening physically. It describes how different observers observe the world, but, in the context of a physical description, these different views contradict each other. Furthermore, it's unlikely that space physically morphs into time and vice versa. Taking Special Relativity as directly describing what's happening physically leads to many, well known paradoxes. In contrast, Lorentz Relativity gives a single, coherent, consistent, direct description of what's happening physically. Physics should be interested in a direct description of what's happening physically.
2) There are phenomena where the physical world that underlies the world of observations peaks through. Two examples, namely, the Twin Paradox and GPS, involve <u>accumulated proper</u> time.
3) Advances in physics would be assisted if physics had an accurate, direct description of what's happening physically.
4) Relativists claim that Special Relativity must be true because the data matches its predictions. If one were to be consistent, then one would conclude that Lorentz clock retardation must be true.
CONCLUSION: A serious analysis of the Twin Paradox and related phenomena call out for a study of preferred frame theory and a more careful analysis of existing data.

A summary of our findings, as detailed in the <u>Twin Paradox Report</u>, is as follows. We investigated published books and journal papers and interviewed proponents of claimed resolutions of the problem. The proposed resolutions were divided into the following categories:
- Resolutions that claimed to employ only the postulates and methods of the special or restricted theory of relativity in which the differential aging effect is due to relative motion.
- Resolutions that invoke the general theory of relativity and, by implication, contend that there is no solution possible from the principles of the special theory of relativity.
- Resolutions that invoke, often implicitly, different or additional assumptions than used in either the special or general theory and, by implication, contend that there is no resolution possible using either theory. Hence, it's clear that there are many conflicting opinions about

the resolution of the Twin Paradox among "mainstream", relativist professors.
Although the mainstream consensus is that the paradox is not a problem and as such has a definitive solution, there is no agreement as to exactly what that solution is as the physics journals and textbooks are full of conflicting solutions to this problem. Hence, we suggest that an open, public discussion of this problem be undertaken with the objective of resolving this critical problem. We ask that, as step one, the "mainstream" physics community select a single, definitive solution to this problem. In addition, we ask that it state which alternative solutions are essentially equivalent to the chosen solution and which alternatives are deemed invalid. If the Twin Paradox is well understood and if there is a generally accepted solution, then this should be a very easy task.

Signed:
Highlighted names are linked to related web pages
Nick Percival, Twin Paradox Project Leader; CNPS (USA)
Harry Ricker, Relativity Group Leader: CNPS (USA)
Greg Volk, CNPS (USA)
David de Hilster, CNPS (USA)
Prof. Dr. Hartwig Thim, Johannes Kepler University Linz, Austria, IEEE, CNPS/NPA (Austria)
Prof. Joe Nahhas, CNPS/NPA (USA)
Robert de Hilster, CNPS/NPA (USA)
Prof. Diego Saá, Escuela Politécnica Nacional, CNPS/NPA (Ecuador)
MA(EE) John-Erik Persson, CNPS/NPA (Sweden)
Francis Viren Fernandes, Scientist, Mediclone Biotech Pvt. Ltd, CNPS/NPA (India)
Prof. Dr. Rati Ram Sharma, Albert Schweitzer Prize-1989 & Nominee for Nobel Prize in Medicine-1996, Postgraduate Institute of Medical Education & Research, CNPS/NPA (India)
Pharis E. Williams, New Mexico Tech (retired), CNPS/NPA (USA)
Ian Montgomery, CNPS/NPA (Australia)
Bruce Harvey, Theoretical Physicist, CNPS/NPA (UK)
Prof. Ian McCausland, University of Toronto, CNPS/NPA (Canada)
Kenneth L. Moore, CNPS/NPA (USA)
Dr. Marvin E. Kirsh, California State University Los

Angeles, CNPS/NPA(USA)
Dr. Pal Asija, CNPS/NPA, Our Pal LLC (USA)
Dr. Janusz D. Laski, CNPS/NPA (Poland)
Dr. Thomas E. Phipps, Jr, CNPS/NPA (USA)
John Remington Graham, CNPS/NPA (USA)
Thierry J. De Mees, CNPS/NPA (Belgium)
Prof. Gary Johnson, Kansas State University, CNPS/NPA (USA)
Charles E. Weber, CNPS/NPA (USA)
Dr. Glenn Borchardt, Progressive Science Institute, CNPS/NPA (USA)
Dr. Michael H. Brill, CNPS/NPA (USA)
Ronald R. Hatch, 24 GPS patents, Institute of Navigation, CNPS/NPA (USA)
S. I. Wells, CNPS/NPA (USA)
Ravil B.Kalmykov, CNPS/NPA (Russia)
Prof. J. G. Klyushin, Saint-Petersburg University of Civil Aviation, CNPS/NPA (Russia)
Prof. Stephan J.G. Gift, The University of the West Indies (Trinidad and Tobago)
Viraj Fernando, CNPS/NPA (Canada/Sri Lanka)
Prof. Franco Selleri, Istituto Nazionale di Fisica Nucleare, New York Academy of Sciences, Fondation Louis de Broglie, Telesio Galilei Academy of Science, Bari University, CNPS/NPA, past Board of Directors of the Italian Physical Society-CERN-Saclay-Cornell-Dubna (Italy)
Jocelyne Lopez, CNPS/NPA (Germany)
Prof. Frederic Lassiaille, Polytechsophia, CNPS/NPA (France)
Dr. Jeremy Dunning-Davies, Department of Physics(retd), University of Hull; President Telesio-Galilei Academy of Science, CNPS/NPA England)
Dipl. Ing. Ekkehard Friebe, CNPS/NPA (Germany)
Dipl. Ing. Peter Ripota, CNPS/NPA (Germany)
Dipl. Phys. Hans Deyssenroth, CNPS/NPA (Germany)
Prof.(retd) Johann Marinsek, CNPS/NPA (Austria)
Prof. Zifeng Li, Yanshan University CNPS/NPA (China)
Karlheinz Baumgartl, Kosmologe (Germany)
Seadin Jelovac, CNPS/NPA (Bosnia and Herzegovina)
Reiner Bergner, CNPS/NPA (Germany)
Lin Haibing (China)
Prof Dr Velimir Abramovich, University of Belgrade, CNPS/NPA

(Serbia)
Rothwell Bronrowan (Germany)
Bai Tao (China)
Wei Enqing (China)
Ph.D Peter Kohut, CNPS/NPA (Slovakia)
Dr. Niels v. Festenberg, Technische Universität Dresden, DPG (Germany)
Dr. Wolfgang Engelhardt, Max-Planck-Institut für Plasmaphysik, CNPS/NPA (Germany)
Dr. Peter Hayes, University of Sunderland, CNPS/NPA (UK)
Robert L. Henderson, CNPS/NPA (USA)
Dr. Józef Kajfosz, CNPS/NPA (Poland)
W. H. Owen, CNPS/NPA (Australia)
Xinwei Huang (China)
Dr. Sergey N. Arteha, Space Research Institute, CNPS/NPA (Russia)
Walter Babin, Editor, General Science Journal, CNPS/NPA (Canada)
Dionysios G. Raftopoulos, Dipl. Mech.-EE of NTUA, CNPS/NPA (Greece)
Prof. Dr. Tolga Yarman, Okan University, CNPS/NPA (Turkey)
Dr. Egbert Scheunemann, CNPS/NPA (Germany)
Dr. Joseph Levy , CNPS/NPA (France)
Barrie J Tonkinson, IEE, CNPS/NPA (UK)
V. N. Kochetkov, FSUE, CNPS/NPA, (Russia).
Dr. D. F. Roscoe, School of Mathematics (retd) University of Sheffield, CNPS/NPA (UK)
G. O. Mueller, CNPS/NPA (Germany)
Dr. Milos Abadzic, CNPS/NPA (Serbia)
Ing.(grad.) Lothar Pernes, CNPS/NPA (Germany)
John Doan, CNPS/NPA, JDX (Australia)
Dr.-Ing. Wolfgang Lange (Germany)
Henry Lindner, CNPS/NPA (USA)
Dipl. Chem. Ing. Herbert Sommer Resalt, (Spain)
Stephen J. Crothers, CNPS/NPA (Australia)
Nainan K. Varghese, CNPS/NPA (India)
Dr. Luigi Romano, CNPS/NPA (Italy)
Parwis Nabavi (Germany)
Associate Professor Boon Leong Lan, Monash University, CNPS/NPA (Malaysia)
William Gaede , Researcher @ ViNi, CNPS/NPA (Argentina)

Notfinity Process

Karl Reiter (Austria)
Dr.-Ing. Günter Dinglinger, CNPS/NPA (Germany)
Dr. Michael Harder, Bureau of Interdisciplinary Sciences (Germany)
Luitpold Mayr, CNPS/NPA (Germany)
Dr. rer. nat. Zycha Harald (Austria)
Dr. Phys. Silvano Lorenzoni (Italy)
Ing. HF Gerhard Klose, DARC (Germany)
Prof.(em) Alfred Evert, CNPS/NPA (Germany)
Mustafa Sprecic, CNPS/NPA (Bosna i Hercegovina)
Dipl.-Ing. Dr. Wigbert Winkler (Austria)
Dr. Ajay Sharma, Shimla Gaurav Awardee, FPS, CNPS/NPA, (India)
Dipl.Math. Johannes Rasper, CNPS/NPA (Germany)
Alexander Weise (Germany)
Dr. Nikolay Chavarga, CNPS/NPA (Ukraine)
Associate Prof. Yang Xintie, NPU (China)
Harald Hölbling (Sweden)
Theophanes E. Raptis, DAT -NCSR Demokritos, CNPS/NPA (Greece)
HSG/USG Peter Herzig (Switzerland)
Eduard Bardas, (Germany)
Dr. Manfred Lichtinger (Germany)
Ricardo V. Consiglio (Brazil)
José Miguel Ledesma, Engineer (Argentina)
Joseph A. Rybczyk, Independent Researcher, CNPS/NPA (USA)
Joachim Blechle (Germany)
Azzam AlMosallami, The Science Center for Studies and Research (Palestine)
Eyüp Firat, Theoretical Physicist, General Science Journal (Turkey)
Dipl. Ing. Alexandar Nikolov, Technical University, Sofia; CNPS/NPA (Bulgaria)
Dr. Graeme Heald, NATA (Australia)
Antonio Saraiva, CNPS/NPA (Portugal)
Antonis N. Agathangelidis, CNPS/NPA (Greece)
Eng. Alfredo Dimas Moreira Garcia (Brasil)
Christos A.Tsolkas, (Greece)
Prof. Bernard Guy, Ecole N.S. des Mines de Saint Etienne (France)
Dipl.-Ing. Claus Grüning, Schule (Germany)
Ing. Christian Sutterlin, Jean de Climont associates Ltd, CNPS/NPA, (France)
Denys Lépinard, WSM (France)

Absurd QMT, SRT and GRT

Dr. Yefim Bakman, Tel Aviv University (Israel)
Ron Heath (USA)
Prof. Bertrand Wong, Eurotech, S'pore Branch (Singapore)
Peter G. Bass, CNPS/NPA (U.K.)
Nillo Gallindo (Brazil)
Steve Waterman, CNPS/NPA (Canada)
Paul Talbot (Canada)
Harald W. Sommer (Austria)
Mamoru Hidaka (Japan)
Roger J Anderton, CNPS/NPA (UK)
Eng. Jesus Sanchez (Spain)
D. Birks (USA)
Marcus Coleman (Australia)
Ing. Nicolae Joica (Romania)
Dr. Nico Benschop, Amspade Research (Netherlands)
Dr. Craig Dilworth, Uppsala University (Sweden)
Prof. Sonu Kumar (UK)
Prof. Sitthichai Pookaiyaudom, Mahanakorn University of Technology (Thailand)
Prof. Milan Kecman, CNPS/NPA (Bosna i Hercegovina)
Prof. Viv Pope, POAMS (Wales)
Prof.(ret.) Dr. Johan F. Prins, Sage Wise 66 (Pty) Ltd (South Africa)
Ingenieur Dane Gacesa (Serbia)
Dr. Jan Meijer (Netherlands)
Dr. C. Johan Masreliez, EST Foundation (USA)
Luiz Ernesto Credidio Mura, IPEN (Brazil)
Maciej Rybick (Poland)
Mohammad Shafiq Khan (India)
Forrest Bishop, CNPS/NPA (USA)
Prof. Florentin Smarandache (USA)
Dr. Al McDowell (USA)
Dr. S. V. Shevchenko (Ukraine)
Glenn A. Baxter, Prof. Eng. (Physicist), CNPS/NPA, The Scientific Journal (USA)
Dr. Victor Kuligin, Voronezh State University (Russia)
Dipl.-Ing Christoph W. Weritz, DARC (Germany)
Andrew Laidlaw (Australia)
Dipl.Ing Branko Vasiljev (Croatia)
Ivan Fordjarini (Serbia)

Dr. Bill McCann, SUST (China)
Tom Hollings (England)
Dipl. Ing. Octavian Balaci (Romania)
Dr. Anastasia-Maria Leventi-Peetz (Germany)
Msc.Eng. Andrew Wutke (Australia)
Dr. Johan Masreliez (USA)
Dipl. Ing. Dimiter G. Stoinov CNPS/NPA (Bulgaria)
Dr. Emiro Díez Saldarriaga (Columbia)
George Coyne Director CNPS (Canada)

Implications for Cosmology and Physics

Because the basic assumptions of SRT are wrong, the theory is fundamentally flawed, which means that cosmology and Einstein's relativity have an invalid foundation. In asserting that SRT directly describes what's occurring physically, such as time dilation depicting different proper time accumulation rates, then contradictions such as the Twin Paradox occur, this refers to the idea that if an identical twin travels into space at very high speed, each twin is moving with respect to the other, but upon returning home the Earth bound twin will have aged more.

The data contradicts GRT so fudge factors such as dark matter and dark energy were devised to try to fix the discrepancy, which is so enormous that now cosmologists claim that 95.1% of the universe is comprised of these. To account for the data the abstraction of inflation was concocted to support the BBT. The fact that this required the cosmos to increase by a factor greater than 10^{78} in less than 10^{-32} seconds does not concern its supporters.

Establishment's Intolerance of Dissenting Views

The brilliant plasma physicist and electrical engineer Hannes Alfvén is a prominent example of someone who scientific journals rejected submissions from, for the above-described reasons. Alfvén is the recipient of the 1970 physics Nobel Prize for magnetohydrodynamics work and the prestigious Bowie medal of the American Geophysical Union for his solar system work. His space science research led to applications in magnetosphere studies, Van Allen belt theory, and formation of comet tails, reduction of Earth's magnetic field in magnetic storms, development of solar system, physical cosmology, and galactic plasma dynamics. Contributions to astrophysics were on identified nonthermal synchrotron radiation from cosmic sources, and on the galactic magnetic field. Alfvén was very significant in the development of solar phenomena research, plasma physics, interplanetary medium, charged particle beams, aurorae science, magnetohydrodynamics and magnetosphereic physics. Technologies using his work include rocket propulsion, hypersonic flight, controlled thermonuclear fusion, particle accelerators, and space vehicles reentry braking. Scientists named Alfven's waves for him.

One could logically expect that such an accomplished scientist would not meet resistance in getting papers published in prestigious journals. However, mostly only obscure ones, which most scientists could not access easily, agreed to publish them. American journal *Terrestrial Magnetism and Atmospheric Electricity* declined his theoretical paper on magnetic storms and auroras because it differed with orthodox physics.

References

[1] Borchardt, *The Scientific Worldview*.344.

[2] Ibid.

[3] Heisenberg, *Physics and Philosophy, the Revolution in Modern Science, 67* .

[4] Heisenberg, *Physics and Philosophy, the Revolution in Modern Science, 67.*

[5] Marmet, Paul citing Heisenberg in *Absurdities in Modern Physics: A Solution Or: A Rational Interpretation of Modern Physics.* Ch 4-2.

[6] Heisenberg, *Physics and Philosophy, the Revolution in Modern Science*, 45.

[7] Ibid.

[8] Weizsäcker, Carl Friedrich von, quoted by Cramer, in *The Transactional Interpretation of Quantum Mechanics*,77.

[9] Jones quoting Cramer in *What Is the Copenhagen Interpretation of quantum Mechanics?*

[10] Ibid.

[11] Feynman, *The Strange Theory of Light and Matter.*
[12] Ibid.
[13] Heisenberg, *Physics and Philosophy, the Revolution in Modern Science*, 43.
[14] Ibid, 49.
[15] Norsen and Nelson, *Yet Another Snapshot of Foundational Attitudes Toward Quantum Mechanics.*
[16] Mermin, Is the Moon There when Nobody Looks? Reality and the Quantum Theory, in *Physics Today*, 47.
[17] Borchardt, *Is Space Matter.*
[18] *Howell, How Many Stars Are In The Universe?*
[19] The Physics Hypertextbook, *Motion.*
[20] Nappi, B. in e-mail comment to me.
[21] Zitzewitz, Paul, Neff, Robert F., Davids. Mark, *Physics Principles and Problems, 725.*
[22] Wikipedia. Field (physics).
[23] Merriam-Webster, *Definition of Property.*
[24] Science Direct. *Electromagnetic radiation.*
[25] Einstein, Introduction by Nigel Calder, *Relativity The Special and the General Theory*, 45.
[26] Puetz and Borchardt, *Universal Cycle Theory, 12.*
[27] Borchardt, *Do spent batteries have less mass?*
[28] Daniels and Alberny, cited in *The Scientific Worldview*, 61.
[29] Borchardt, *The Scientific Worldview*, 62.
[30] Nappi, B. in e-mail comment to me.
[31] Borchardt, *The Ten Assumptions of Science*, 10, 125.
[32] Borchardt, *The Physical Meaning of $E=mc^2$.*
[33] Ibid.
[34] *Borchardt, Einstein's most important philosophical error.*
[35] Khamehchi, Hossain, Mossman, Busch, McNeil Forbes, and Engel, *Negative-Mass Hydrodynamics in a Spin-Orbit–Coupled Bose-Einstein Condensates.*
[36] BBC Science and Environment, *Physicists Observe "Negative Mass."*

[37] Nappi, B. in e-mail comment to me.
[38] Borchardt, Negative mass?
[39] Albert Einstein, *Relativity The Special and the General Theory*,69, 70.
[40] Ibid.
[41] Borchardt, *Infinite Universe Theory.*
[42] Bethell, Tom, Rethinking Relativity
[43] Manor, E. P. *Gravity, Not Mass Increases with Velocity.*
[44] Ibid.
[45] Ibid.
[46] Santilli, R., *Nine Theorems of Inconsistency in GRT with Resolutions via Isogravitation.*
[47] Van Flandern, T., The Speed of Gravity—What the Experiments Say.
[48] Bethell, Tom. Rethinking Relativity.
[49] Ibid.
[50] Ibid.
[51] Einstein, A, Explanation of the Perihelion Motion of Mercury from General Relativity Theory.
[52] Bethell, Tom, quoting Van Flandern in Rethinking Relativity.
[53] Dingle, H. and O' Keefe, M. B., Science At the Crossroads.
[54] Bethell, Tom. Rethinking Relativity.
[55] Percival, N., *Twin Paradox, Data Does Not Match Special Relativity Time Dilation*
[56] Ibid.
[57] Ibid.
[58] Ibid.
[59] Ibid.
[60] Ibid.
[61] Ibid.
[62] Ibid.
[63] Ibid.
[64] Ibid.
[65] Ibid.
[66] Ibid.
[67] Percival, N., *An Open Letter to the Physics Community Twin Paradox.*

Part Four **Theories from Observations**

Chapter 6

Modern Mechanics Replaces Relativity

Until 2007, attempts to falsify relativity theory by questioning its assumptions, experimental results and non-intuitive implications, have not succeeded because its defenders have used the following three established arguments: They claim that challengers do not understand it properly, the predictive results are supportive, and lack of intuitiveness is irrelevant to the theory's validity. Objections based on its subtle math errors were insufficient to succeed because of the predictive capability of its equations and workable proofs. However, because of Steven Bryant's critical examination of the math in relativity, this is no longer the case.

From examining Einstein's spherical wave proof, Bryant found no such wave formed in the moving frame, as Einstein indicated, but instead a transformed sphere or ellipsoid was present.[1].

In a sphere or spherical wave all points are equidistant from a common fixed point. However, Einstein's crucial mathematical mistake invalidates the derivations. As Bryant discovered, Einstein neglected to treat the right-hand side of the equation as unchanging or constant in the transformed equation. Therefore, the transformed shape is not a spherical wave, nor do the lengths indicated by the equations have the same length or originate at the shape's center.[2] Thus Bryant has reached a valid conclusion that Einstein's spherical wave claims fail. This means the proof he uses to establish relativity also fails. Relativity theory does not follow from its assumptions because the sequence of accepted statements from one result to the next one contains important errors in math.[3]

Because relativity theory derives from this proof, it is essential, and its failure, which is a failure to develop a second sphere, eliminates relativity's basis. Without knowing where the problem is in the work, the failed proof indicates that the there is definitely something wrong in it.[4]

Bryant addresses the widely held erroneous belief that the proof passes. He is successful in challenging Einstein's derivations. He demonstrates that the proof is not passable because it has a Type I error, which involves an incorrect rejection of a true null hypothesis. This means that the relationship that Einstein supposedly established between the crucial speed of light constancy hypotheses, and the principle of relativity, is missing. As Bryant points out this "false positive" has implications to Einstein's abstractions of time dilatation, curved spacetime and length contraction.[5]

Steven Bryant provides the following clarification at the stevenbbryant.com website: "My main argument is against spherical wave proof that Einstein provides in Section 3 of his paper that established relativity theory: "Zur Electodynamik bewegter Koerper" (On the Electrodynamics of Moving Bodies).Ultimately, that paper is the primary one that we must consider in evaluating the validity of Einstein's theory. With Steve Bryant's permission, I am including this English language translation of the proof from *Disruptive: Rewriting the Rules of Physics*.[6]

Image 3. Einstein's "Spherical *Wave Proof*"

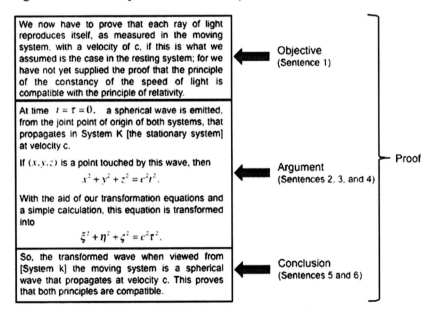

Bryant states: "Einstein says that he must show that his two principles are associated and compatible. He offers his spherical wave proof as evidence of that relationship. Einstein believes that the second shape is a spherical wave (Sentence 5), which enables him to conclude that relativity is valid (Sentence 6). I have shown that Einstein's fifth sentence is not tru—that a spherical wave is not the transformed shape, which means that his sixth sentence is not true. The spherical wave requires that, in addition to maintaining the second equation's equality (which is the only characteristic most people consider), each segment must be the same length and originate from the transformed shape's center to its surface. Neither of the later conditions are met."[7]

General Relativity Invalidated

GRT predicts that as distances from a star increase the bending of light decreases, although even when further out than many times the star's radius, relativity predicts the star's gravitation will still bend light. Contradicting this, Edward Dowdye Jr.[8] showed that beyond the plasma atmosphere, there is no bending and therefore the curved space abstraction is incorrect. Microwaves and light along a path of minimum energy or least time deflect only in the Sun's plasma rim[9] because of refraction.

For many years, Glenn Borchardt has written on how Einstein's relativity does not adhere to `The Ten Assumptions of Science,'[10] a fact that unequivocally means that Einstein's model does not meet the threshold that is required to be considered scientific.

In his 2016 book, *Disruptive*, Bryant presents the proof that SRT and GRT are invalid due to their uncorrectable mathematical mistakes, and he shows that Einstein's $E=mc^2$ formula is only an approximation.[11]

In addition to the very serious errors of confusing and misusing types and the failure of Einstein's spherical wave proof, Bryant reveals several major problems with relativity. In the crucial initial step of his theory formulation, Einstein was unsuccessful in an accurate derivation of a spherical cone,[12] which is an expanding sphere from a fixed-point source. He commits a critical error in portraying relativity as a two-system model when it is actually a non-nested three-system model with an oscillating system in the ray of light, a moving inner-system and an outer stationary system.[13]

Bryant provides detailed discussion of Einstein's significant errors. He explains one of these in which Einstein came to the illogical conclusion that there is no need to apply the same Newtonian equations used throughout his work once the formal proof is completed.[14] This refers to his derivation that states that the inner-system's movement accords with the translation equation $x=x^1+vt$, and then ignores this at the derivation's conclusion.[15]

As Bryant reveals, relativity develops from Einstein not correctly understanding the Tau, and his treatment of it. Einstein incorrectly assumed that in moving systems Tau designates time, when in fact, the Tau function results in an average (intercept) time for an oscillating ray of light to reach an inner-system point and back in a non-nested relationship.[16] Bryant comments that based on this, Einstein's x, y, and z equations are average intercept lengths from the origin to three inner-system points of the inner-system, not coordinates of a particular inner-system point.[17]

Bryant points out that Einstein's faulty understanding of the Tau equations and functions resulted in him treating average intercept lengths as inner-system specific point coordinates. Einstein invokes the Tau function, known as "Einstein's time adjustment" without first defining the function body, which means he does not know what the Tau is or does.[18] (Invoking it without defining it first is common practice in mathematics). In defining his Tau function, Einstein used a partial differential equation (PDE). However, doing this led to an error because PDEs result in functions, not equations.[19] (For explanation of functions, refer to the Appendix 1 `Functions used in MMT`).

Einstein's equations include three serious problems for which no fixes are possible, because as stated Einstein incorrectly describes relativity as a two-system model attempting to explain behavior of three-system.[20] One problem is that the equations do not provide an accurate description of relativity because they are average intercept lengths and times (i.e. nadals) to three points of the moving system.[21]

Another problem resulted because of representing the position of the moving system; Einstein incorrectly normalized the average intercept lengths, not realizing that the translation transformation must apply to the moving system. Not acknowledging this normalization led to the third problem, which is the accuracy level of the equations.[22] Einstein provided no analysis of his Tau.[23]

The Michelson-Morley experiment is a problem for Einstein's equations because when examined through a statistical lens, they do not properly account for all behaviors associated with the electromagnetic force, something Modern Mechanics Theory (MMT) does very easily. In addition, relativity's accuracy does not exceed MMT's and produces some inferior results.[24]

By showing relativity's serious problems that cannot be resolved, Bryant invalidates it, thereby removing the fundamental support for the theory. Because relativity appears to have considerable experimental support, and performs very well, although not quite as accurate in predictions as MMT, many physicists have enormous difficulty accepting that it is invalid. Their resistance, even when confronted with the mathematical evidence provided by Bryant, is likely due to years of indoctrination into this model. After graduation, they treat relativity theory as axiomatic and many go on to teach it at universities or write about it. When physicists come to identify with it, they are taking the most harmful approach that a scientist can have. In doing science, one must always avoid this trap and never treat any theory as sacrosanct.

Einstein's Errors

For many years, Glenn Borchardt has written on how Einstein's relativity does not adhere to `The Ten Assumptions of Science,`[25] a fact that unequivocally means that Einstein's model does not meet the threshold that is required to be considered scientific.

In his 2016 book, *Disruptive*, Bryant presents the proof that SRT and GRT are invalid due to their uncorrectable mathematical mistakes, and he shows that Einstein's $E=mc^2$ formula is only an approximation.[26]

In addition to the very serious errors of confusing and misusing types and the failure of Einstein's spherical wave proof, Bryant reveals several major problems with relativity. In the crucial initial step of his theory formulation, Einstein was unsuccessful in an accurate derivation of a spherical cone,[27] which is an expanding sphere from a fixed-point source. He commits a critical error in portraying relativity as a two-system model when it is actually a non-nested three-system model with an oscillating system in the ray of light, a moving inner-system and an outer stationary system.[28]

Bryant provides detailed discussion of Einstein's significant errors. He explains one of these in which Einstein came to the illogical conclusion that there is no need to apply the same Newtonian equations used throughout his work once the formal proof is completed.[29] This refers to his derivation that states that the inner-system's movement accords with the translation equation $x=x^1+vt$, and then ignores this at the derivation's conclusion.[30]

As Bryant reveals, relativity develops from Einstein not correctly understanding the Tau, and his treatment of it. Einstein incorrectly assumed that in moving systems Tau designates time, when in fact, the Tau function results in an average (intercept) time for an oscillating ray of light to reach an inner-system point and back in a non-nested relationship.[31] Bryant comments that based on this, Einstein's x, y, and z equations are average intercept lengths from the origin to three inner-system points of the inner-system, not coordinates of a particular inner-system point.[32]

Bryant points out that Einstein's faulty understanding of the Tau equations and functions resulted in him treating average intercept lengths as inner-system specific point coordinates. Einstein invokes the Tau function, known as "Einstein's time adjustment" without first defining the function body, which means he does not know what the Tau is or does.[33] (Invoking it without defining it first is common practice in mathematics). In defining his Tau function, Einstein used a partial differential equation (PDE). However, doing this led to an error because PDEs result in functions, not equations.[34] (For explanation of functions, refer to the Appendix 1 `Functions used in MMT`).

Einstein's equations include three serious problems for which no fixes are possible, because as stated Einstein incorrectly describes relativity as a two-system model attempting to explain behavior of three-system.[35] One problem is that the equations do not provide an accurate description of relativity because they are average intercept lengths and times (i.e. nadals) to three points of the moving system.[36]

Another problem resulted because of representing the position of the moving system; Einstein incorrectly normalized the average intercept lengths, not realizing that the translation transformation must apply to the moving system. Not acknowledging this normalization led to the third problem, which is the accuracy level of the equations.[37] Einstein provided no analysis of his Tau.[38]

The Michelson-Morley experiment is a problem for Einstein's equations because when examined through a statistical lens, they do not properly account for all behaviors associated with the electromagnetic force, something Modern Mechanics Theory (MMT) does very easily. In addition, relativity's accuracy does not exceed MMT's and produces some inferior results.[39]

By showing relativity's serious problems that cannot be resolved, Bryant invalidates it, thereby removing the fundamental support for the theory. Because relativity appears to have considerable experimental support, and performs very well, although not quite as accurate in predictions as MMT, many physicists have enormous difficulty accepting that it is invalid. Their resistance, even when confronted with the mathematical evidence provided by Bryant, is likely due to years of indoctrination into this model. After graduation, they treat relativity theory as axiomatic and many go on to teach it at universities or write about it. When physicists come to identify with it, they are taking the most harmful approach that a scientist can have. In doing science, one must always avoid this trap and never treat any theory as sacrosanct.

Motion in MMT

MMT explains motion using the math associated with geometric transformation. MMT equations account for the fact that it is a three-system model, having two moving systems: a bidirectional oscillating one (e.g. a vibration, an alternating current or a swinging pendulum) and an inner system. It also has an outer system[40]

Distance, duration, and position equations describe the mathematical relationship between systems. In addition to referring to static length between points, MMT makes use of dynamic distance equations to express the effort needed to move between positions, to describe motion, and for how far a system moves relative to its outer-system from an origin and a specified point in relation to time and velocity.[41] In three-system models, they describe an oscillating system's behavior. Duration refers to measured time, which I call "nadal." Position equations represent a system's location in reference to its outer-system. Modern Mechanics uses geometric transformation. This involves mapping from one plane to a second one to perform positional changes of a system. Bryant explains that to change a system's position, MMT uses translation transformation, by adding a distance to its original position to find its new position.[42] (Translation preserves orientation in addition to congruence.) The position equation describes a system's motion in reference to time (i.e. nadal), length and velocity.

Bryant explains that the MMT model describes its interactions using nine equations referring to intercepts for positions, lengths and times. MMT defines the relationship between the inner-system and an oscillating system in relation to distance, time and space.[43] MMT uses distance and surface equations to describe wave motion, and distance and position equations for particles' motion.[44] The intercept time function provides the average, reflected and forward intercept times; the intercept length function does the same for length.[45]

Ives-Stillwell

In 1938, Herbert Ives and G.R. Stillwell designed this to test relativistic time dilation to the Doppler shift of light. Primarily H_2^+ H_3^+ ions in combination accelerate through perforated plates charged from 6,788 to 18,350 volts. The researchers wanted to measure two things: the Doppler displacement of a hydrogen atom in a contained ray tube, and the shift in half a wavelength.

Image 4. Ives-Stillwell experiment[46]

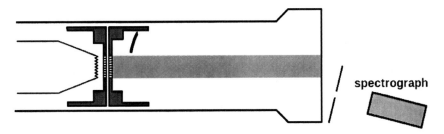

Using a concave mirror offset at 7 degrees, experimenters observe the beam and its reflection. The result agreed with the transverse Doppler effect formula and appeared to support the concept of time dilation. The experiment used a construction to show that Einstein's SRT equations accurately predict "time dilation," Supporters of relativity theory used the results as evidence for this absurd abstraction, which is dependent on imaginary spacetime. Additionally relativity supporters claimed it to be proof for Einstein's energy equations and SRT because it was the only theory able to account for the 1938 Ives-Stillwell[47] experiment.

Although physicist Georges Sagnac had disproved the idea of time dilation, this did not dissuade relativity supporters. However, once Bryant released his precursor to Modern Mechanics, the Model of Complete and Incomplete Coordinate Systems in 2007, an accurate alternative explanation to SRT became available. MMT predicts a Doppler shift mean wavelength of 15.69, which is the same as the observed 15.69. SRT predicted 15.72.[48] Thus, MMT is more accurate than SRT in explaining the Ive's-Stillwell experiment.

Bryant's Reassessment of Michelson-Morley Experiment

Michelson and Morley[49] designed their 1887 experiment to measure the change in wavelength of light waves as the Earth revolves around the Sun. They expected to be able to measure changes, which they would use in determining Earth's orbital velocity by comparing the speed of light in perpendicular directions through the hypothesized luminiferous stationary "aether wind." Assuming that aether moves with respect to the Sun, they thought that their measurements would equal Earth's orbital speed of 30 kilometers per second. Instead, after applying their equations to the raw data, they arrived at only 8 kilometers per second for the orbital velocity of Earth. From this result, they concluded that the experiment is incompatible with completely or partially dragged aether models. The experiment had tremendous significance because it appeared to invalidate all aether theories.

Bryant provides clear, logical analysis of this experiment, whose data when properly interpreted support the existence of a model off partially dragged aether.[50] One of the problems that Bryant discovered was that the interferometer could not measure length and time, but Michelson and Morley's equations were only properly applicable to discrete types, such as distance and time[51]. He points out that analyzing the raw results with wrong equations ensured that the experiment would have failed regardless of the presence of aether.[52] Another error Bryant discovered that the two researchers made was their wrong assumption that a fringe width is one wavelength rather than the correct one-half of a wavelength.[53] The final problem was how the experimenters operated the interferometer and how it performed.

There was less than one chance in 1,000 that the calculations of the raw data indicated a null result. Bryant used this data with the correct math to determine a velocity of 32 kilometers per second. This corresponds to what the researchers expected to find where there is dragged aether involved.[54]

When Bryant analyzed and interpreted this experiment correctly, he found that it produced a result of 30 kilometers per second for Earth's orbital velocity, which validates the partially dragged aether theory[55] The significance of this discovery to relativity physics is immense.

Speed of Light and Fine-Structure Constant

Since 1998 credible evidence has been accumulating that the "fine structure constant," also known as alpha is not really a constant at all. Puetz and Borchardt's Neomechanical Theory proposes that alpha varies over time.[56] If this is the case, it has major significance for relativity.

Alpha relates to the electromagnetic coupling constant that involves bonding strength of an elementary charged subatomic particle with an electromagnetic field. It measures how strongly these particles interact with light and one another. It is a collection of three measurements, which are supposed to be unassailable as constants. One number, which is 1/137.0359, generates from Planck's constant, the speed of light, and the charge of an electron. Physicists refer to this number as a dimensionless constant, because it does not refer to any specific quantity amount, such as seconds.

Because the speed of light is inversely proportional to alpha, if alpha is not actually a constant, this would contradict a fundamental principle of SRT that the speed of light is invariable over time and distance. It also conflicts with GRT"s equivalence principle relating to gravity and inertial mass. If gravity and quantum mechanics equations are not consistent over space and time, or if those equations involve a change in numbers that are supposed to be constants, then it would indicate that the most fundamental laws of physics have undergone changes over time.

Solid evidence exists that 2 billion years ago on Earth the speed of light, one of the most hallowed of the physical constants in physics and cosmology, and indispensable to special relativity theory, was slower than it is today.[57] Changes in alpha were found by Victor Flambaum and associates at the University of New South Wales in Sydney in 2004. This was determined by analyzing relative concentrations of radioactive isotopes left behind at Oklo in a natural nuclear reactor, which began almost two billion years ago in Gabon, a nation on the west coast of Central Africa,

In 1998 astrophysicists, John Webb and Victor Flambaum with their associates at the University of New South Wales commenced a search for evidence of variations in alpha. Their study included hundreds of northern-sky quasars. These are extremely luminous and distant galaxies powered by massive black holes at their centers and inferred to be over 10 billion light years away. Webb and Flambaum examined light from the quasar's active galactic cores. Absorption by gas clouds en route results in some of the quasars' wavelengths, or colors, of light coming toward us ceasing to be part of the spectrum.

Interactions between atoms in the gas and photons from the quasar light result in absorption of wavelengths, which indicate the cloud's chemical composition. The absorption spectrum has the fine structure that provides the value of alpha. This enables astrophysicists to detect changes to this assumed constant over many billions of years by examining the spectrum of light from the quasars. The fine-structure constant reveals differences between wavelengths absorbed by any two elements.

In August 2001, the researchers published their results, which included an analysis of 13 potential sources of systematic error. They found that none of these could account for the increase in alpha which they concluded was smaller 6 billion years ago by approximately one part in 100,000.[58] This study confirmed their previous work.

Webb and his colleagues made additional observations[59] in 2010. They observed more than 100 quasars. The team used the Very Large Telescope in Chile and the W.M. Keck Observatory in Hawaii, in order to cover the southern and northern skies. Their results indicated that during much earlier eras of the Cosmos, alpha had a different value. They concluded it was about one part in 100,000 smaller than it is today in the northern sky and one part in 100,000 larger in the southern direction. The model's data have a one in 15,000 chance of being a random result [60]

In September 2010, Webb presented their results to the Joint European and National Astronomy Meeting in Lisbon, Portugal. In addition, the researchers submitted their work to the journal *Physical Review Letters* for peer review.

After re-analyzing, the data in March 2004 S.K. Lamoreaux and J.R. Torgerson determined that a small increase in the speed of light had occurred. The basis for this finding[61] was that alpha has decreased by more than 4.5 parts in 100 million since Oklo was active. It is notable that not a single scientist has reported any problems in their assumptions and analysis.

These discoveries regarding changes to the fine-structure constant violate a supposed axiom of SRT regarding the absolute constancy of the light's speed, and they do not conform to GRT's equivalence principle between inertial mass and gravitation. Confirmation of these results will bring a big change to physics and cosmology. It may be sufficient to invalidate both relativity theories.

Electromagnetic Force

Whereas relativity gives no defined result for objects moving at the speed of light, MMT describes objects that are below that speed, those that match it, and hypothesized ones over the speed of light. MMT is capable of describing multiple wave mediums, such as water, air, and aether that could transmit at slower or faster velocity than the speed of light in a vacuum. Bryant contends that MMT might be suitable to describe undetected wave mediums with faster- than-light propagation.

References

---

[1] Bryant, *Disruptive,* 13-29.

[2] Bryant, e-mail to me.

[3] Ibid., 14 -29.

[4] Ibid., 30.

[5] Ibid., 13-29.

[6] *Disruptive Rewriting the Rules of Physics.*

[7] Bryant, *Feedback- Responding to Questions and Com*ments.

[8] Dowdye Jr. *Significant Findings.*

[9] Ibid.
[10] Borchardt, *The Ten Assumptions of Science.*
[11] Ibid, 271.
[12] Ibid., 17, 26.
[13] Ibid., 281.
[14] Ibid 282.
[15] Ibid.282.
[16] Ibid., 282.
[17] Ibid., 282.
[18] Ibid., 162.
[19] Ibid., 162.
[20] Ibid, 281.
[21] Bryant, *Disruptive,* 202.
[22] Ibid. 162.
[23] Ibid., 203.
[24] Bryant, *Disruptive*, 251.
[25] Borchardt, *The Ten Assumptions of Science.*
[26] Ibid, 271.
[27] Ibid., 17, 26.
[28] Ibid., 281.
[29] Ibid 282.
[30] Ibid.282.
[31] Ibid., 282.
[32] Ibid., 282.
[33] Ibid., 162.
[34] Ibid., 162.
[35] Ibid, 281.
[36] Bryant, *Disruptive,* 202.
[37] Ibid. 162.
[38] Ibid., 203.
[39] Bryant, *Disruptive*, 251.
[40] Ibid., 85.
[41] Ibid, 93.
[42] Ibid, 99.

[43] Ibid., 102.
[44] Ibid ,120.
[45] Ibid., 120.
[46] Stigmatella aurantiaca, **Ives**-Stillwell experimen
[47] Ibid., 251.
[48] Bryant, *Disruptive,* .253.
[49] Michelson, & Morley,1887, *On the relative motion of the earth & the luminiferous ether.*
[50] Bryant, *Disruptive*, 216.
[51] Ibid, 223.
[52] Ibid, 223.
[53] Ibid, 230.
[54] Ibid, 31-35.
[55] Ibid, 238.
[56] Puetz and Borchardt, *Universal Cycle Theory*, 213.
[57] Flambaum and Dzuba, Kozlov, Angstmann, .Berengut, Cheng, Karshenboim, Chin, Nevsky, Porsev, Ong, Derevianko and others, *Effects of variation of fundamental constants and violation of symmetries (P, T,) in atoms July*9,2016.
[58] Berg, *Focus: A Constant that Isn't Constant.*
[59] *Johnston, Changes spotted in fundamental constant.*
[60] *Ibid.*
[61] Lamoreaux and Torgerson, *Neutron moderation in the Oklo natural reactor and the time variation of α.*

Chapter 7

QMT vs. Alternatives

Eighteenth century philosopher Paul-Henri Thiry, Baron d'Holbach, stated:

> "The universe, that vast assemblage of every thing that exists, presents only matter and motion: the whole offers to our contemplation, nothing but an immense, an uninterrupted succession of causes and effects."[1]

Compare that view of the Universe with the quantum physicists' conception of a world that has no actual events, but has probabilities for action, which require someone to perform a measurement for it to become an event in our world.[2]

QMT Contradictions

The contradictions in QMT would likely not have been a problem for Joseph Stalin, the former dictator of the Soviet Union, based on this statement of his:

> "We are for the withering away of the state, and at the same time we stand for the strengthening of the dictatorship of the proletariat, which represents the most powerful and mighty of all forms of the state which have existed up to the present day. The highest development of the power of the state, with the object of preparing the conditions of the withering away of the state: that is the Marxist formula. Is it 'contradictory'? Yes, it is 'contradictory.' But this contradiction is a living thing and wholly reflects the

Marxist dialectic."[3]

Compare Stalin's statement to the QMT approach to contradictions, whose theorists embrace them. A good example is the one inherent in the wave-particle duality concept, in which it is claimed that the behavior of objects as large as molecules have features of both particles and waves, Niels Bohr, its principle originator, regarded this 'duality paradox' as a fundamental reality. His principle of complementarity stated that wave and particle behavior are mutually exclusive, and it is not possible to observe or measure simultaneously their complementary properties. The principle states that to describe phenomena completely we need to analyze them separately for particular aspects, such as behaving as a stream of particles or as a wave.[4]

Similar to Stalin, rather than realizing that contradictions reveal a profound failure of a theory, they accept the contradiction, thinking incorrectly that it reflects what exists.

Matter

Glenn Borchardt provides a good definition for matter as "an abstraction for the world of physical objects. It is the name that we give to the class that includes all things."[5] These objects, also known as microcosms, always contain other microcosms. So even if it were hypothetically possible, which it is not, that everything in the category of matter had an ultimate tiniest fundamental unit, in a completely solid indivisible form with no components or space within it, it could not be considered to be matter. On the assumption that only forms of matter can exist, it is impossible for non-material to exist, just as it is for "nothingness" to exist. This is because the concept of "existence" only applies to things. It does not refer to motion, because that occurs rather than exists. Additionally, based on the assumption that to exist requires that it be matter, if there is no matter present, then there is no existence. Thus, an ultimate building block, which would have to be non-material, cannot exist.

There is no logical reason to think matter can have any existence without some empty space, but there is plenty of evidence that particles involve far more space in their makeup than solidness. For example, atoms are 99.999999999 percent empty of ordinary baryonic matter, although that apparently empty area would have an enormous amount of aether and sub-ether *ad infinitum*. I have more discussion about matter in Chapter 8's Theories without Contradictions.

It is not theoretically possible for matter to go down to absolute zero degrees, although physicists have reduced the temperature of particles to .000000000101°K.[6] Even at that record low temperature, there is still internal motion. Many people mistakenly believe that if we could get an object to absolute zero, then all molecular motion would cease. In Alan Lightman's *Great Ideas In Physics* (2000), he appears to give credence to this view in stating that a substance reaches absolute zero if all of its molecules are motionless except for ordered motions.[7] Although he adds the last two words as a qualifier, the statement could leave the impression that molecules can be in a motionless state. It would have been more precise if he had clarified that even at the hypothetical temperature of absolute zero, molecules would still vibrate with zero-point energy and "translational motion," even though no heat energy from molecular motion is available to transfer.[8]

In addition to this internal motion, all microcosms are moving because their position is still changing relative to Earth's rotation and orbit of the Sun, which travels through the galaxy. The galaxy is also moving through space, etc. Thus, every object's position relative to the rest of the Universe is constantly changing, which translates to moving. The abstraction of motionless matter implies all internal motion of a microcosm to stop and all matter in the Universe to cease all movement. Because this cannot occur, the idea of matter without motion is absurd. The other point to consider is that without any internal quantum level motion, it is impossible for a microcosm to exist. Matter exists because of its motion.

QMT Duality of Light Concept

The concept of wave-particle duality discussed in Chapter 5 as a preposterous concept in which the phenomenon of light supposedly exists as both a particle and as a wave, contradicts basic logical thinking. I am amazed that so many physicists accept the idea that comes from the QMT interpretation of the two-slit experiment. This indicates that in the education of physicists, courses in critical thinking need to be mandatory. That would soon result in a new generation of physicists who reject

Superposition and Probability Waves

Another irrational abstraction applied by QMT advocates in trying to explain many impossible assertions is "superposition." In this concept, quantum systems, such as an atom or a "photon," correspond to various potential outcomes. This continues until there is an interaction with something external or an observation is made of it, causing its collapse into one of the definite states that were previously just possibilities. According to this fundamental principle of QMT, things supposedly exist in every possible state simultaneously. Quantum physicists claim that interference peaks from an electron wave in a double-slit experiment are an observable manifestation of superposition.

Image 5. Double-slit physics experiment[9]

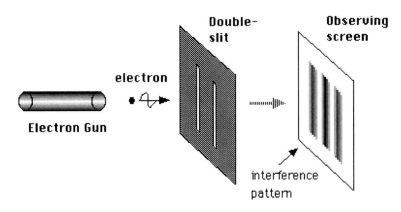

Superposition is an absurd way to explain the apparent behavior of light. In the double-slit experiment, a laser beam, or other coherent source of light or electrons, illumines a barrier, such as a plate, with one slit open, then with both slits open. After going through the slits, it contacts the observation plane. When two slits are used, particles including electrons and photons create a wave pattern.

QMT Contradictions: Interpretation of Double-slit Experiment

QMT physicists believe that particles and light have properties consistent with Newtonian particles and waves. Their explanation is that light manifests as photons and that individual photons travel on two paths simultaneously toward two detectors until one of them detects a photon.

Image 6. Two slits illuminated by a plane wave[10]

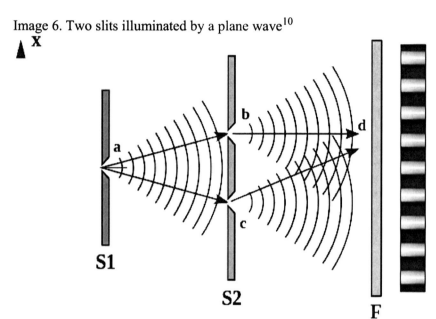

To avoid explaining how a wave can exist without a material involved that is waving; physicists have redefined "wave" to refer to a description of properties. Thus, in QMT wave simply designates the presence of propagations, frequencies, interference, and properties that are usually associated with a wave.

QMT needed this new meaning for waves to seem to invalidate the perfectly reasonable question of what the wave is composed of in the vacuum state. This way of defining waves also can appear to remove relevance to the behavior of other quantum mechanical systems.

In the two-slit experiment when both slits are open with transparent detectors registering which slit a photon passes through, two photons never go through both openings simultaneously. However, without detectors, the result indicates that two light waves go through the two slits at the same time, just as is expected for a spread-out wave, which we see in the interference pattern on the photographic plate. These contradictory experiments present the mystery of what is occurring in the double-slit experiment.

The QMT explanation for these results is that the researcher's choice of which of the two experiments to conduct determines the outcome for whether light shows particle or wave behavior, making the observer an integral part of the experimental system.

The double-slit experiment is very significant for seeming to demonstrate the wave-particle duality of matter, and because it apparently negates the reality of a universe without our observations. That is the interpretation made by QMT proponents.

The experiment led many physicists to ponder many fundamental questions about reality.[11] These include ones about the role performed by the observer and whether observing the world can be separated from what is observed. For those who believe in the reality of photons, there is the question about whether they are a mixture of particle and wave between emission and detection and how that is possible. The question arises over how to characterize matter prior to its observation. Physicists have also considered whether observation creates the Universe.

QMT physicists believe that before observations occur, it is never even theoretically possible to describe the properties and the existence of anything with more definiteness than the probabilities associated with it. The QMT position is that it is not correct to refer to existing objects prior to making a measurement of them. The fact that they believe there are only probabilities, demonstrates the incredible level of absurdities that these physicists accept in supporting a QMT interpretation of the double-slit experiment as opposed to the de Broglie–Bohm Pilot Wave Theory.

If a light wave is wide enough to travel both routes, a ray tracing will correctly predict maxima and minima on the screen when several particles produce the expected interference pattern after going through the apparatus. Adherents of the QMT model are only interested in the equipment, reactions and mathematical formulae that provide information of what transpires at an atomic scale. They use this to make accurate experimental predictions. To do this they construct a probability wave with crests and troughs to indicate probabilities. Although this is merely a product of thought, it purportedly and miraculously exits the mind and travels through space as if it is a material wave. Because QMT physicists believe mathematics is equivalent to reality, they accept this nonsense as being valid.

This model precludes knowing a particle or photon's location until detection occurs. The photon has to be in a superposition, which means all positions, throughout the observation path. After their emission, to get an interference pattern, particles must be able to take two or more paths to a detector screen.

For QMT to be correct several impossibilities must occur. Part of a light particle must pass through one slit, and then another part of it enters through the other slit. For the resulting interference pattern to appear, the part that enters a slit first must modify its course in the secondary field after supposedly deciding on a destination on the detector screen. Next, it must get this information to the other particle so that part could change its course to reach the screen simultaneously with the other part in the same location. We are expected to accept as a given that photons are capable of making decisions.

QMT has multiple problems involving the error of confusing abstractions with something real. Physicists mistakenly treat probability waves, which are a pure mathematical concept, as if they are real and can pass through space. Superposition is an abstraction that is only representing an imaginary impossible world.

Communication from Physicist Basil Hiley

In this August 27, 2014 e-mail to me Basil Hiley talks about his work and a proposed experiment. It is described in their paper *Weak measurement and its experimental realization,*[12] where they discuss the relationship between the real part of the weak value of the momentum operator at a post selected position. They re-examine the significance of the experimentally determined stream-lines in the Kocsis et al Toronto experiment.[13] Their paper contends that energy flow lines are not "photon trajectories." They propose doing an analogous experiment using the Bohm approach which predicts well-defined trajectories of atoms, to directly compare the two-slit trajectories of a beam of atoms calculated by Philippidis, Dewdney and Hiley.

QMT versus Alternatives

Dear George,

I have made considerable progress on the formal side of his ideas showing how the theory comes out of an algebraic approach to QM started by von Neumann. This involved showing how the approach can be rigorously formalised in terms of symplectic and orthogonal Clifford algebras. In the way we can handle spin and relativity. I was hoping that this work would show that the Bohm approach is not just philosophic speculation, but lies at the heart of QM shedding more light on quantum phenomena. But it now turns out that when I try to explain these ideas to physicists their eyes glaze over, clearly uneasy with the mathematics even though they use it implicitly. However they then retreat into their own take on QM ignoring what we have done.

To try another approach, I have teamed up with a couple of experimental physicists at UCL and we have just been awarded a grant of US$670K to perform some experiments to give the Bohm approach to QM some experimental backing. You can find out what we are trying to do in a paper Flack, R. and Hiley, B. J., Weak Measurement and its Experimental Realisation, {\em J.Phys:Conference Series}, {\bf 504} (2014) 012016. If we are successful then I hope people will wake up and take notice.

Best wishes,
Basil.

On this same topic in a June 9, 2017 e-mail he provides further clarification on his earlier one in this excerpt: The Heisenberg uncertainty principle asserts you cannot measure position and momentum simultaneously. This means you cannot experimentally tell whether the particle has or has not got a simultaneous position and momentum. The standard approach assumes the particle does not have a simultaneous position and momentum and works out the consequences. The Bohm approach assumes it does have a simultaneous position and momentum and then works out the consequences. One of the consequences is that the QP appears which is absent in classical processes. Thank you for the Facebook url re-Penrose.

Best wishes,
Basil.

Steven Bryant's MMT Interpretation

In *Disruptive*, Bryant[14] uses the deterministic realist de Broglie-Bohm Pilot Wave Theory to describe how waves in the aether are similar to light waves, which are a type of motion produced by pilot waves. In MMT, the term "objects" replaces the words "waves" and "particles". Bryant makes a comparison to a boat going through water creating bow waves moving outward from the front, which travel ahead of the boat.[15] He explains that objects in the double-slit experiment generate similar interference patterns[16] as the ones generated by mutually reinforcing multiple wave fronts. He mentions that several intersecting waves passing through the slits create a single pattern on the observation plane.[17]

Applying the Pilot Wave Theory, Bryant explains the experiment by referring to a pilot wave, produced by an object moving through the primary field, which goes faster than that object. Bryant points out that these waves cannot change an object's direction in the field after going through a one-slit plate, because they move away from the object. He clarifies that in the two-slit situation, waves emitted from both open slits into the secondary field move at higher velocity than the objects, forming an interference pattern in that field.[18]

After going through a slit into an active secondary field, an object undergoes a directional impact. Once reaching a propelled peak, it presumably rides to the observation plane. Bryant clarifies that what is mistakenly referred to as an "interference pattern" on the observation plane is actually just valleys forming due to both waves being 180 degrees out of phase at the destination, while peaks result from the waves being in phase, which refers to waves from both sources reaching the observation plane simultaneously.[19] His explanation of the experiment removes the need for the irrational abstraction of particle-wave duality. I find the pilot wave solution to be understandable and credible, in sharp contrast to QMT's bizarre explanation of the double-slit experiment. Additionally, because it requires that we make only the assumption of a pilot wave, compared with the multiple impossible assumptions of quantum mechanics, MMT is superior to the QMT model based on Occam's razor. This is the principle that simplest solutions are the most likely ones to be correct.

Schrödinger's cat Thought Experiment

Erwin Schrödinger, one of the two principle creators of QMT, attempted to reveal absurdities of the Copenhagen interpretation of QMT with a thought experiment known since then as Schrödinger's cat. In this scenario, a steel container has a Geiger counter, which is out of the cat's reach, with a bit of radioactive substance, sufficiently tiny that over one hour the possibility of a single atom decaying is 50 percent. Such an occurrence will cause the counter tube to discharge and via a relay, it releases a hammer that breaks a container flask of hydrocyanic acid. After one hour, we can state that if no atom has decayed, the cat is alive. The system's psi-function expresses this by having in it a dead cat and a living cat smeared or mixed out in equal parts.

In cases such as these, it is usual that atomic domain indeterminacy transforms into indeterminacy in the macroscopic world, which resolves through observation. This stops us from accepting a "blurred model" as a valid representation of reality. Schrödinger mentioned that by itself it would not contain any contradiction or be unclear; and that a difference exists between a snapshot of clouds and fog banks and an out-of-focus photo [20]

Rather than being deterred by this powerful argument by Schrödinger against the Copenhagen interpretation of QMT, its proponents have embraced the experiment whole-heartedly as literally representing reality. They evaluate other interpretations of QMT based on how these deal with this experiment. Richard Feynman won the 1965 Nobel Prize in physics and the British journal Physics World placed him amongst the 10 greatest physicists ever in a poll of 130 top physicists in the world in 1999. In reference to the paradoxes in QMT, such as Schrödinger cat being both simultaneously alive and dead, the best answer he could generate was that the "paradox" is only a conflict between reality and your feeling of what reality should be.[21]

To those who saw that the QMT Copenhagen interpretation made no sense, his response was: "I think I can safely say that nobody understands quantum mechanics."[22] He expressed the view that to think you understand QMT indicates that you do not.[23] QMT physicist John Wheeler stated: "If you are not completely confused by quantum mechanics, you do not understand it."[24] If this is the best defense of quantum theory by the top theoretical physicists, then the theory must be weak and seriously flawed, in which case we need to discard it immediately. Einstein stated his position very well in contending: "The universe is real—observing it does not bring it into existence by crystallizing vague probabilities." Explaining why he could not accept quantum mechanics, he commented that he likes to think that the moon exists even when he is not looking at it.

Misinterpreting Entanglement Phenomena

Physicists who adhere to the ideas of QMT accept that quantum entanglement (QE) as depicted in their model is a reality. QE occurs when two quantum systems, such as particles or groups of particles, link in such a way that their linear momenta in one direction and spatial coordinates in one direction have a 1:1 relationship. QE is evident in a correlation for properties, such as polarization, position, up or down spin, and momentum of paired particles propelled in opposing directions from a single source. Borchardt writes:

> "The combined systems can be described as a whole unit, sort of like the ends of a dumbbell. Therefore, ascertaining the momentum or position for a single quantum system will result immediately in setting those properties for the paired one. In the case of the dumbbell we can see the connection between the two ends, but in quantum entanglement, that is not so clear."[25]

Although there is no argument over the correlation, the QMT explanation is preposterous because it requires particles to not only be somehow communicating with one another, but also doing this in zero time.

In order for the experiment to prove quantum entanglement, the researchers would have to adhere to the following protocol: A laser beam fired through a certain type of crystal splits individual photons into pairs of entangled photons (photon A and photon B). Physicists would follow these photons, separated by any distance, during the entire experiment. After doing thousands of measurements to photon A and noting corresponding changes to photon B's spin (e.g. from an up spin to a down spin and back to an up spin etc), regardless of how many observations are made every time the physicist measures specific photon A's spin, then the specific photon B's spin will have the opposite spin. Unless this is the experimental protocol, then the results are not meaningful.

When I described this scenario to Glenn Borchardt, his response was:

"First of all, there are no such things as `photons.` At best, many of the properties attributed to photons are the properties of individual aether particles. I doubt that any of what you mentioned is other than mathematical imagination. I have speculated that there are 10^20 aether particles in an electron (from Planck's constant and equations), so I can't imagine determining the spin of any one of them, much less a pair of them. BTW: I would love to have someone show me an experiment that does what you suggest. I don't need the math, just the data that proves it. In my opinion, data is far more significant as a proof than math."[26]

The QMT explanation for entanglement is absurd. Attempting to make observations conform to QM theory results in a misinterpretation of what is happening and is then labeled quantum entanglement. Duncan Shaw's excellent paper regarding polarization of aether provides a good example of this. I discuss this idea further on in this section.

Bill Westmiller[27] came up with the following interesting take on QE. As entanglement experiments always involve conservation of matter in motion, one should anticipate that paired objects would have reversed spin to one another. Westmiller points out that this is because they emit from one source and therefore must have reciprocal properties and directions. In addition, he comments that the instruments are only able to measure positive or negative spin regardless of the particle's orientation, which can be as much as 90 degrees from perpendicular. Westmiller concludes that considering these facts, one should not give any significance to experimental results that are within instrumentation error rates.

For the paired particles to be somehow "communicating" their quantum state to one another (which requires accepting his bizarre notion) instantly irrespective of galactic distances would require faster than light transmission of information. As this violates Einstein's universal speed limit, Einstein rejected QE as requiring "spooky action at a distance." He explains his objection in a 1935 paper that he wrote with Boris Podolsky, and Nathan Rosen, *Can Quantum-Mechanical Description of Physical Reality Be Considered Complete?* Now known as EPR, the paper contends that the wave function cannot sustain completeness of the quantum description without violating the principle of locality.

Bell's theorem is the counter argument to EPR. It states that no physical theory of local hidden variables can ever reproduce all of the predictions of QMT. John Bell's 1964 paper, *On the Einstein-Podolsky-Rosen Paradox*[28] provided an analogy to the EPR paradox. He said a measurement decision concerning one paired particle should not influence its paired particle. However, by using a math formulation based on realism and locality, Bell provided cases where this would not equal QMT predictions.[29]

Clauser and Freedman (1972),[30] and Aspect and others (1981),[31] showed that in this respect, QMT is accurate. Although their experiments apparently proved Bell-inequality violations, thereby excluding all local hidden variable explanations for QMT, they do not do this for non-local hidden variables. However, more importantly, as mentioned previously, the experiment employed such a seriously flawed protocol that the results are without any significance.

Bell provided the following non-QMT explanation for entanglement involving absolute determinism:

> "Suppose the world is super-deterministic, with not just inanimate nature running on behind-the-scenes clockwork, but with our behavior, including our belief that we are free to choose to do one experiment rather than another, absolutely predetermined, including the decision by the experimenter to carry out one set of measurements rather than another, the difficulty disappears. There is no need for a faster-than-light signal to tell particle A what measurement has been carried out on particle B, because the universe, including particle A, already 'knows' what that measurement, and its outcome, will be."[32]

Although absolute determinism, as an explanation for Bell's theorem, does not require bringing in the nonsensical superposition of QMT particles, it is inappropriate in many ways and it requires a finite universe. As Glenn Borchardt informed me: "It cannot work because the correct analysis is based on UD which is based on infinite causality instead of finite causality."[33] UD refers to univironmental determinism, which states that whatever happens to a portion of the Universe, referred to as a "microcosm", depends on the infinite matter in motion within that microcosm and the infinite matter in motion external to that microcosm.

Bell's absolute determinism explanation requires that a particle is capable of knowing something. Bohm theorized that this was possible because an electron has a "protomind" that deciphers information about the situation it is in using the quantum potential's (QP) form. In this theory, the protomind does not occur because of the work of neurons, but rather it is present because the electron is part of the whole, which Bohm hypothesized to have a mind.

A powerful argument against Bell's theorem can be made by challenging the validity of the assumption that math is all that we need to represent reality. Bell's theorem supposedly proved that those opposing entanglement were wrong. However, in fact the test was incapable of doing this in the real world, because it was restricted to Set Theory and Venn Diagrams, which are pure basic forms of Math.

Using his method, the experimenters created a set of inequalities, which they employed to set limits. If real word violations occur, then the experimenter falsely declared quantum entanglement theory correct and the realist opponents wrong. The major problem was the incorrect assumption that equations are the essence of reality. It is evident that Bell's inequalities, based on his methodology, are only applicable to formal logic and purely formal relations, which are imaginary.

Rather than accepting the absurd idea of superposition of particles, or Bell's absolute determinism, it seems more reasonable to deny the concept that QE is causing the observations in the experiments. Borchardt addresses this point in this statement:

> "Again, phenomena that display `action at a distance` are `spooky` only to aether deniers. Without aether, we are stuck with 'curved empty space,' 'curved spacetime,' or the magical `attractive force` that still makes no sense even though it has been a solipsistic favorite for centuries. What seems to be `action at a distance` is most likely a local effect produced by variations in aether pressure, as we suggested as the neomechanical cause of gravitation."[34]

Physicists have no way to know what state either particle of the entangled pair was in before the measurement. Those subscribing to the QE abstraction, claim that the paired particles were in a "superposition." This illogical concept is that all things throughout the Universe are in many states until someone observes, or a device takes a measurement, of a thing that then causes it to be in just a single or a smaller set of states, thereby ending superposition.

In QE, superposition supposedly ends only in the measuring of one of them, thereby causing the particle's superposition to collapse into one particular state. This nonsense violates all reason and common sense. Its intended purpose is to support the concept of quantum entanglement.

The QE abstraction is a perfect illustration of a concept that developed because of confusion between a formal imaginary mathematical description and what could possibly occur in the Universe. Most importantly, it demonstrates how physicists come to accept the most implausible and even impossible theories because of not recognizing when they are dealing with an abstraction that has no association with reality.

Quantum physicists have decided on one particular interpretation of entanglement experiments, and have not been willing to consider other descriptions. This is unfortunate. When I was in my 20s, I thought that I did not understand something about the concept of quantum entanglement, because it made no sense to me. However, as I got older it became clear to me that the QMT physicists do not comprehend the facts.

Bohm and Hiley's Entanglement Explanation

An atom at its lowest possible energy level is in its ground state. In David Bohm and Basil Hiley's[35] understanding, it has entirely quantum potential energy (QP), which they describe as an information potential that guides quantum particles' movement. They point out that we cannot lever it out without affecting conservation.

By including and focusing on QP, entanglement takes on a new aspect. QP's implication for quantum non-locality determines the significance of entanglement. The form of the QP energy in entanglement is the explanation for the ascension of non-locality.[36] This description sees particles and photons moving through space as an inseparable part of the pilot wave.

Threshold Model

Eric S. Reiter clarifies how in Newtonian and QM models a particle means something different, and so does a wave. In classical physics a particle is anything that stays together unlike waves that are incapable of this. However, particles in QM are a mixture: non-physical quantum waves of probability can disperse everywhere and then collapse to a particle. His paper[37] describes his particle-like and wave-like experiments, with matter using alpha rays, and light using gamma rays. Reiter states that the two-for one effect he obtained only looks that way if you use the standard QM interpretation.[38] He assesses the results of these beam-split coincidence experiments as offering support for his model, which embraces a pre-loaded sub-quantum state." Reiter concludes: Entanglement is an illusion of the threshold and ratio properties of charge, action and mass.[39] He welcomes others to observe or adjust his experiments and to replicate them.

Shaw's Elegant Solution to Entanglement

A theory that does not involve particles traveling through space is Duncan Shaw's very plausible, intuitive solution. He argues against the fundamental mistake and rationale of considering entanglement to be a QMT phenomenon of particles or photons moving through space to a receptor. Disputing QMT explanations for phenomena believed to represent QE, he contends that what occurs are waves transmitting through aether. His rational explanation derives from seeing the so-called spin correlations as polarization of Maxwell's pervasive aether.[40] In a 2012 paper, Shaw recommended that physicists reconsider aether as being a subatomic wave for electromagnetism.[41] As Shaw writes, Maxwell theorized that when electromotive force caused stress for aether, the outcome was polarization.[42] In referring to several experiments that investigated entanglement, he shows the significance of physicists' QMT assumptions leading them away from any consideration that emissions were waves propagating through aether, which could explain the correlated results in properties.[43]

Shaw provides a very illustrative quote by Maxwell describing aether's motion:

"Thus, then, we are led to the conception of a complicated mechanism capable of a vast variety of motion, but at the same time so connected that the motion of one part depends, according to definite relations, on the motion of other parts, these motions being communicated by forces arising from the relative displacement of the connected parts, in virtue of their elasticity."[44]

Shaw contends that waves of electromotive force reaching a target activate the aether parts, referred to as "cells" at that place, which appears as if photons have arrived.[45]

In the quantum physicists' QE assumption, polarization refers to properties, such as momentum of photons going to the target. However, in Maxwell's model, polarization is the aether forced state resulting from electromotive force applied to it.[46]

Electromotive force polarizes aether, diffused throughout all of space, which makes aether form 3-D patterns that offer polarization planes for the transmission of electromagnetic waves. The rotation of these planes of polarization can result in the electromagnetic waves rotating.[47][48]

Electromotive energy disperses in opposing directions from a single source causing the polarization patterns to be a mirror image in the aether medium in both directions. This is due to the electromotive energy going out in opposite directions from one source. That also explains why the double-slit experiment observation screen shows reverse data from the arriving waves.[49]

The action-at-a-distance illusion vanishes by the proposed existence of aether cells interacting with particles, which determines correlated results.[50]

Because rotating waves are hitting receptors, the readings taken are only of rotations of the electromagnetic waves, although to the physicists these represent up or down spin. It is clear that QM theorists are mistaken in thinking that the observing screen has any role in entanglement phenomena other than to record data[51]

I concur with Borchardt's assessment of Shaw's entanglement paper. In an e-mail Glenn sent to me, he stated: "The best thing on quantum entanglement is Shaw's recent paper."[52] Shaw's entanglement solution is easily understandable, consistent, and without contradictions or so-called paradoxes. Significantly, it does not require accepting ideas that violate rationality or logic. QMT entanglement explanations cannot make these claims. If his explanation of entanglement phenomena holds up, and it gets widespread acceptance by physicists, the scientific impact will be profound. Without a perceived foundation in quantum entanglement, QMT will cease to be of any significance in creating models of reality. Then science, unshackled by the constraints of QMT, will be able to make huge advances. No longer will publications reject papers, making potentially valuable research contributions, simply on the basis that they do not conform to QMT. How many scientific papers exist in this category is unknown. The work of the following brilliant plasma physicist and electrical engineer is one prominent example.

Ridiculous Declaration that Universe is Impossible

Following the reputed discovery of the Higgs particle, researchers at King's College London (KGC) combined observations made with the Background Imaging of Cosmic Extragalactic Polarization (BICEP) telescope with current understanding of the Higgs particle. After careful consideration of prevailing astronomical theories, they concluded that if these theories are accurate, it is "a massive impossibility" for the Universe to exist based on accepted models of particle physics and the Universe's origin.[53]

Although one may contend an infinite universe is an abstraction because we cannot refer to its defining boundaries, nevertheless there are very real things, such as stars, planets, and particles, that most definitely exist. However, according to the KGC scientists none of the things covered by the category of matter should exist if present orthodox physics is correct. As things obviously do exist, the only rational conclusion is that the physics theories do not represent reality.

Rather than discard the theories that result in clearly absurd conclusions, astrophysicists seem willing to accept the ridiculous proposition that the Universe's existence is not possible. The fact that these researchers use "massive" with "impossible" reveals that they have little comprehension of what the word "impossible" means, because it is absurd to have greater degrees of impossibility. However, considering that QMT physicists believe many impossible things, it is understandable that they would continue believing in the orthodox models, while accepting the ludicrous idea that the Universe cannot possibly exist.

Just as illogical as the KGC scientists' contention, is the QM concept that quantum particles can exist simultaneously in more than one location. Does this make any rational sense to anyone other than true believers of the standard QMT QM model? It does not only have the appearance of being unscientific; it is in fact pure pseudoscience.

Some physicists try to deflect any attention from the absurdity of their belief system by confidently claiming that the Universe is very weird. No sane person would try the same tactic in a court of law if charged with a crime. Imagine anyone telling a judge that even though the evidence against him or her was enormous, the judge should accept his or her implausible testimony as true. Would anyone claim innocent with the explanation that it is just due to the nature of the Universe that makes it seem that his or her guilt is obvious? Someone would only try such a tactic if attempting to make a case for an insanity plea.

Since Herman Weyl's (1919) and later Einstein's attempts (1921) to unify these two theories, physicists have continued working toward this goal of creating one grand theory in which both of these would be subsumed. To date there has been no success in unification of these confused theories, riddled with contradictions and absurdities. Attempting to do this has no likelihood of producing a greater understanding of the Universe.

Even the most ardent proponents of QMT are unable to offer any rational reason to accept its descriptions of what supposedly occurs. That is why they always rationalize it on the basis that it is the most predictive theory. However, with the advent of Modern Mechanics, with a predictive ability that exceeds relativity,[54] there is now a credible challenger to QMT in the prediction arena.

Richard Feynman attempted to address the absurdities of QMT with a warning against posing the obvious questions: "Do not keep saying to yourself, if you can possible avoid it, 'But how can it be like that?' because you will get 'down the drain', into a blind alley from which nobody has escaped. Nobody knows how it can be like that."[55]

How can a theory be considered scientific if it has so many profound absurdities that one will go "into a blind alley" if questioning how its logically impossible claims can be possible?

Those in authority positions in the physics and cosmology community who are proponents of these two models are very much like the Emperor in Hans Christian Anderson's allegorical *The Emperor's New Clo*thes. In this hilarious fable, a very narcissistic emperor is only concerned with showing his subjects how wonderful he looks in his clothes. His new weavers promise to make him the best suit using a fabric that will be invisible to "hopelessly stupid" people or those not fit for their jobs. As the weavers are making the suit, the emperor and his ministers cannot see it, but they pretend they can so that others will not think them to be unfit for their positions.

When the suit is completed, the vain emperor marches in procession.

Everyone maintains the pretense, out of fear of appearing stupid or unfit for his or her positions. After a while, a child yells: "But he isn't wearing anything at all!" Others in the crowd repeat this, which the emperor hears. Even though he knows this is correct, he continues the procession, keeping up the pretense. Science needs more people like that child to expose contradictory orthodox theories based on impossible and nonsensical abstractions.

In a June 7, 2017 blog,[56] Glenn Borchardt writes how a similar situation to the foolish emperor exists within the scientific establishment today: "Remember that in science, a paradigm is a set of theories, experiments, and interpretations that are used to advance a particular discipline. A mature paradigm sponsors what Kuhn called 'ordinary' science. As in religion, disloyalty can result in rejection or banishment. Unlike religion, it seldom results in imprisonment or execution (with rare exceptions such as Galileo and Bruno)."Science needs more people like that child to expose contradictory orthodox theories based on impossible and nonsensical abstractions.

References

[1] Stanford *Encyclopedia of Philosophy, quotes* Baron d'Holbach.

[2] Herbert, *Quantum Reality: Beyond The New Physics,* 27.

[3] Wikipedia cites Evans, quoting Stalin in Soviet Marxism-Leninism: The Decline of an Ideology.

[4] Sleator, *Quantum Mechanics*: *Uncertainty, Complementarity, Discontinuity and Interconnectedness.*

[5] Borchardt, *The Scientific Worldview,* 32.

[6] Aalto University School of Science , *World record in low temperatures.*

[7] Lightman, *Great Ideas In Physics*, 44.

[8] Encyclopedia Britannica, *Absolute Zero.*

[9] Nekoja, Wikimedia, *Double-slit experiment.*

[10] Francesco, Franco, *2 slits illuminated by a plane wave* Wikimedia

[11] Lightman, Great Ideas In Physics, 202.

[12] Hiley, B and Flack, R. Weak measurement and its experimental realization.
[13] Kocis, S., Braverman, B., Ravets, S., Stevens, M. J., Mirin, R. P., Shalm, L.K., Steinberg, M. A., Observing the Average Trajectories of Single Photons in a Two-Slit Interferometer,
[14] Bryant, *Disruptive*, 256-269.
[15] Ibid., 266.
[16] Ibid., 262.
[17] Ibid., 266.
[18] Ibid., 267.
[19] Ibid., 262.
[20] Schrödinger, *The present situation in quantum mechanics, A Translation of Schrodinger's "Cat Paradox Paper."*
[21] Feynman, *The Character of Physical Law - 6 -Probability and Uncertainty.*
[22] Ibid.
[23] Ibid.
[24] *Quantum Quotes*, quoting John Wheeleer.
[25] Coyne quoting Borchardt in The *Myth of Quantum Entanglement*.
[26] Ibid.
[27] Westmiller, *Spooky action at a distance,* The Scientific Worldview.
[28] Bell. On *The Einstein, Podolsky, Rosen Paradox.*
[29] Ibid.
[30] Freedman and Clauser *Experimental Test of Local Hidden-Variable Theories.*
[31] Aspect, Dalibard, and Roger, *Phys. Rev. Lett.* 49, 1804 –
[32] Bell, *Free will.*
[33] Borchardt, in e-mail to me.
[34] Coyne quoting Borchardt, in *The myth of quantum entanglement.*
[35] Bohm and Hiley, The *Undivided Universe,* 35-37.
[36] Hiley, *From the Heisenberg Picture to Bohm: a New Perspective on*

Active Information and its relation to Shannon Information.

[37] Reiter, Eric S. Experiment and Theory removing all that Quantum Photon Wave-Partiicle Duality entanglement Nonsense

[38] Ibid.

[39] Ibid.

[40] Shaw, *Maxwell's Aether: A Solution to Entanglement* Proceedings of the CNPS (2016).

[41] Shaw, *The Cause of Gravity: A Concept*, Phys. Essays 25, 66

[42] Shaw, *Maxwell's Aether: A Solution to Entanglement.*

[43] Ibid., 39.

[44] Ibid., 39.

[45] Ibid.

[46] Ibid.

[47] Ibid.

[48] Ibid.

[49] Ibid.

[50] Shaw, *Reconsidering Maxwell's Aether* .

[51] Ibid.

[52] Borchardt, December 12, 2016 e-mail to me on Shaw's QE paper.

[53] Shaw, Tina, *Tech Times.*

[54] *Bryant, Disruptive,* p 251.

[55] Feynman, *Probability and Uncertainty - the Quantum Mechanical View of Nature.*

[56] Borchardt, G. BS for detecting loyalty.

Chapter 8

Theories without Contradictions

Matter and Motion are Abstractions

The most problematic use for the term matter is in referring to a substance that composes things. This concept of matter started with Democritus 2,500 years ago with his proposal that the basic constituents of everything were indivisible atoms filled with an inert substance referred to as matter. In theoretical physics, matter is not a basic concept. In fact, because the term is always a category for what exists, it is actually an abstraction.

I differentiate between "existing" and "occurring." Existence involves mass and volume. Occurrence involves matter's motion. Thus, using this criterion, fields and EMR occur rather than exist.

We can always define particular objects or specific particles of any size by their mass and volume. If these measures cannot be applied, then not valid to claim the existence of what is hypothesized. That is why QMT and relativity physics are mistaken in postulating point particles such as quarks and leptons, which physicists consider having rest mass but no volume. Nappi makes the following comment about the concept of point particles: "If there was a way for a point-centralized gravitational field to move around—i.e. a graviton—then it would have a rest mass. On the other hand, there is no reason it would have inertia. So Einstein's conjecture of the equivalence of gravitational mass and inertial mass would not follow.

1

Theories Without Contradictions

Attempting to get measurements when applying these mass and volune criteria to matter is infeasible because we cannot ascribe any boundaries or mass to matter. Although every particle and galactic supercluster is an example of matter, it is incorrect to think of matter as being a "thing." Since the late 19th century, physicists have been searching for the imagined ultimate tiniest particle. However, this is attempting to do the impossible, because matter is a category, not a "thing" that is "composed" of particles,

Therefore, matter as a thing does not exist. I avoid using the word "it," when referring to matter, because that would imply an existing "thing." We can accurately use the term only to mean a category containing all things.

Since the late 19th century, physicists have been searching for the imagined ultimate tiniest particle. However, this is attempting to do the impossible, because matter is not composed of particles, just as vegetables are not composed of carrots, potatoes, or spinach etc. It is more accurate to state that the category of vegetables includes those items, but the classification has no actual existence; only particular members of it exist. It is impossible to eat a mental construct called lettuce; we can just eat a specific head of a particular variety of lettuce contained in the lettuce category.

Matter refers to everything that exists, and for the same reason that there is no such "thing" as vegetables, there is no thing as matter, only examples of this.. Therefore, it is just as absurd to seek matter's building blocks, as it is to look for one for vegetables.

Matter is an abstraction for existing things, because it is impossible to refer to its boundaries or mass. Only existing things have these. Even though that which is contained within this category exists, matter is only a thought. We cannot divide this abstraction in a material sense, because it is only possible to subdivide individual things that exist. We can only conceptualize matter into ever more specific categories, such as the plant kingdom, vegetables, carrots, etc.

Because almost everyone has learned to think of matter as a thing, it is not easy to understand that matter is a category. By having this understanding, it becomes clear that it is incorrect to think of matter as "existing." What actually exist are particular microcosms. With this perspective, it is possible to delve further into understanding the Universe.

Even the portion of the Universe that is observable is a microcosm within a larger macrocosm and there is no reason to think that there is an ultimate macrocosm that we can label the "Universe." Every macrocosm is a microcosm in a larger macrocosm ad infinitum.

The idea of infinite cannot represent this, because all concepts are only capable of referring to that which is finite. Therefore, it is useless to try conceiving beyond the capacity of limited thought, because one will inevitably be thinking in terms of the finite. That is why I use the concept of notfinity, which negates the possibility of finity without asserting any replacement for it.

In commenting on motion in *Matter and Motion are Abstractions*, Glenn Borchardt stated:

"Because time is motion, time also is an abstraction and may be used as a substitute for motion. No matter what we call that abstraction, we find that we can only observe and measure specific examples of it. Furthermore: 'motion' or 'time' is relative. Universal time is the motion of all things with respect to all other things in the infinite universe. Because it is impossible to measure universal time, we must settle for individual measurements of the motions of specific microcosms. These all must be done with respect to the motions of still other specific microcosms. By convention, we compare those specific motions with the motion of Earth's rotation on its axis or the motion of the microwave signals generated by atomic clocks when electrons in atoms change levels."[2]

Microcosms within Macrocosms

There is no limit to the pattern of microcosms within macrocosms within a larger macrocosm. Each of us contains microcosms, known as the various organs of the body. These organs contain microcosms, such as cells, which have molecular microcosms, in which there are atoms encompassing protons and neutron microcosms. This process of subsuming microcosms of increasingly smaller sizes continues without end.

To have some understanding of matter it is useful to think in terms of Borchardt's matter-space continuum, which he defines as: "A range or series of microcosms that are slightly different from each other and that exist between what we imagine to be perfectly solid matter and perfectly empty space."[3]

Theories Without Contradictions

The prevalent assumption in physics that things must be composed of smaller things is the basis for the belief that understanding the Universe requires finding out what composes things. This results in the quest for the ultimate tiniest particle. If you take that approach, then how will you ever know that you have found the particle that is so fundamental as to not be composed of anything smaller that future higher technology will uncover?

Scientists used to think that the atom was the ultimate building block of everything until technical developments were sufficient to indicate smaller objects. Proceeding from the orthodox assumption that things are composed of smaller things; does it make sense that such an ultimate smallest item is not composed of anything? Such an item could not be termed matter in the sense that I use this word because the category of matter refers only to microcosms that contain smaller microcosms *ad infinitum*. This is different that the concept that microcosms are composed of particles.

Explaining the logic of the idea for infinite tinier sizes of objects, Borchardt stated:

> "Per the Ninth Assumption of Science, relativism (All things have characteristics that make them similar to all other things as well as characteristics that make them dissimilar to all other things) and its required infinity, no two of the submicrocosms are identical. So we cannot say that any particular type of them is the ultimate tiniest submicrocosm that does not contain an even smaller submicrocosm." [4]

Within all these microcosms, there are specific kinds of motion. In believing that microcosms only exist because of being composed of some type of matter, there would be no need to include various types of motion in defining microcosms. However, it is impossible for a microcosm to exist without motion, as this is impossible due to the law of conservation of matter and its motion.

Motion Required For Microcosms

It is impossible for a microcosm to exist without having some type of motion, because this phenomenon is part of the definition of a microcosm. A motionless microcosm is just as inconceivable as having running without moving. Motion is an occurrence; it is not a thing that exists.

Comprehending how the Universe functions requires understanding that every microcosm, and the macrocosm that surrounds it, are in continual motion. Without motion, neither could exist. However, it is important to understand that motion too, is an abstraction. There are only specific motions pertaining to specific microcosms. Although we can measure these specific motions, the actual motions are not a measurement, because they are occurring independent of measuring them.

We can best understand terms such as momentum and force as descriptions of the motion in microcosms and their motion in relation to their macrocosm. "Energy" refers to the calculation of a microcosm's mass multiplied by the square of its motion. There is as much confusion about the meaning of "energy"`as there is for "time". In understanding energy, one must realize that things exist and are in motion, which occurs. "Energy" `is simply a convenient way to describe the motion of matter, but physicists often mistakenly believe that energy has some sort of existence. This idea is similar to thinking that body mass index (BMI), which is a calculation based on one's height and weight, actually has an existence. To overcome the confusion caused by having objectified the energy concept, we could replace its use with matter-motion terms.

Orthodox physicists use the term energy to refer to an abstract physical quantity that supposedly "exists" in varied forms simultaneously as a property of objects, namely their ability to perform work, and systems that have the capacity contain energy. This ignores the fact that an ability or property cannot be isolated from an object.

When motion is a force associated with an object or objects acting upon other objects, thereby displacing those objects, then the motion of the first object or objects has done work upon other objects. When the motion of matter is only due to other matter outside of it moving, then the sample of matter is not doing work but has the potential to perform work.

Through combustion, gasoline is able to do work. By doing calculations based on engine performance and the amount of gasoline required to move a specific car, one determines that 1 L of gasoline can move the car 8 km. That is the number that refers to the work that the gasoline can do. When the object is in motion, then physicists refer to kinetic energy,` as work done by the object. A problem arises when physicists mistakenly refer to "energy" in terms of an "amount."` Borchardt states the problem with this. "But there is no 'amount of energy' in a system. That is because energy is a calculation and there can be no 'amount' of it."[5]

We use numbers in referring to abilities in many areas. In sports, we ascertain ability through simple mathematics. How well one does in school work and tests depends on how much one apples intellect and academic ability, which results in a specific GPA. We get this through a calculation. However, the GPA does not exist, and does not involve any occurrence. It simply indicates one's ability. Using the same reasoning helps one realize what the calculation called energy shows in specific numbers. Realizing that energy does not exist or occur facilitates a better understanding of this concept. To use a running tiger as an example may help to clarify this. A tiger exists and its rapid movement can be termed running. In physics, we describe this running motion with a calculation known as energy. Thus, energy as a descriptive term refers to the calculation involving the motion of a microcosm.

The preceding definition for energy offers a better understanding of this than the one in use since the 17th century, which is that energy is the ability to do work." This defines energy as a property of objects, transferable to other objects or between the two types, which they conceive as the energy of motion (kinetic) and the 133 energy of location (potential). In this concept of energy as a property, kinetic energy refers to calculations, such as electrical and thermal ones. Potential energy includes chemical. This definition of energy, as the ability or property of matter to do work, can never refer to anything beyond a confusing abstraction. This is because it is impossible to isolate properties from objects; a property considered apart from an object is a meaningless concept. Using the concept of kinetic energy to refer to a "property" of

The preceding definition for energy offers a better understanding of this than the one in use since the 17th century, which is that "energy `is the ability to do work". In that definition, physicists define energy as a property of objects that is transferable to other objects or between the two types, which they conceive as the energy of motion (kinetic) and the energy of location (potential). `In this way of conceiving of energy as a property, kinetic energy` refers to calculations such as those involving electrical and thermal ones. Potential energy includes chemical. By defining "energy" as the ability or property of matter to do work, this definition can never refer to anything beyond a confusing abstraction. Because it is not possible to isolate properties from objects, a property considered apart from an object is a meaningless and impossible concept. Using the concept of kinetic energy to refer to a `property` of matter in motion is wrong. As Borchardt has often pointed out, matter exists and motion occurs, and the word energy is simply a descriptive calculation of matter's motion.

In reviewing this section, Borchardt commented: "I also like to describe behavior as a univironmental interaction. It is not the property of a microcosm, but the interaction between a microcosm and its macrocosm."

Understanding the Heisenberg Uncertainty Principle

In *Dreams of a Final Theory*, Nobel physics laureate Steven Weinberg comments that despite doing every conceivable measurement, only probabilities can be predicted regarding the outcomes of later experiments, although the laws of physics and initial conditions completely determine any physical system's behavior [6]

Basil Hiley and R.J. Callaghan, comment on the meaning of the uncertainty principle in *Delayed-Choice Experiments and the Bohm Approach*.[7] After referring to what it states, about not being able to make simultaneous measurements of a particle's precise momentum and position, they point out that the inability to determine such values concurrently does not imply that for the particle they do not exist at the same time.[8]

There is no experimental way to exclude the possibility that quantum particles lack simultaneous values of position and momentum, because this and the uncertainty principle do not rule it out. It is puzzling to me that so many physicists apparently do not understand this obvious implication of the uncertainty principle. Perhaps they need to develop their logical thinking skills.

Borchardt provides an excellent explanation for why it is not possible to ascertain the position and the motion of a particle simultaneously, which does not infer anything about the reality of these features. Writing at the scientific worldview website on July 27, 2016, he points out that the Heisenberg uncertainty principle is really just an observation that detectors are required to make any measurement on microcosms.[9] Because collisions are required for its operation, any collision that it has with a microcosm results in changes to the microcosm's position, and in decreases or increases in motion.[10] He clarifies that for larger microcosms these changes are of no particular significance and we do not notice them, but it is different for microcosms of the atomic realm, because these are more obviously involved with infinity.

Borchardt states that just because infinity is a feature of reality with an infinite number of microcosms containing infinite submicrocosms, which are aether in motion, this does not indicate that observation is required for reality to exist. It only means that precise detection of a particle's position and momentum at the same time is not possible.

Limitless Microcosms in motion

Applying mass and volume criteria to the phrase "infinite universe" reveals that it refers to an abstraction, because by definition, there are no boundaries and no specific masses associated with such a universe. As there is no ultimate largest structure or complete collection of matter and no volume boundaries, this word cannot refer to a thing or microcosm because all things and microcosms are limited. Universe is a category of unlimited microcosms of an infinite variety of sizes, and the occurrence of their limitless motions and interactions with their macrocosm and with other microcosms. Being an abstraction, the universe does not exist as an identifiable object. This definition removes the idea of applying causation to the Universe because it is merely an abstraction and thus cannot have a cause. Causality only applies to microcosms within this category not to the category. The predominant view of the universe existing as a definable structure creates much confusion in many areas such as "infinity." Although it is impossible to prove whether there are infinite microcosms, or whether other forms of matter not yet discovered exist, Borchardt's infinite universe model, described in Chapter 10, has the advantage of being non-contradictory. This fact alone gives it more credibility than the orthodox alternative with its many unresolved contradictory elements.

It is futile trying to conceptualize "non-existence." which is an impossible and absurd thing to do, because this effort inevitably results in thinking of something. Realizing that it is not feasible to create a model to portray all that exists, with no macro or micro boundaries, shows how limited our theories will always be. Therefore, in discussing "what is," it is important to put the emphasis on falsifying concepts, such as a finite universe, while recognizing the limitations of words and concepts in discussing notfinity

Although we can never fully comprehend the Universe, we can have a better understanding of it by seeing the invalidity of ideas, such as spacetime in relativity, and concepts involving flowing time in traditional western views. western views. Referring to infinite microcosms in motion, rather than "infinite Universe," may offer a better way to comprehend reality. Theory will advance as our abstractions improve.

Notfinity and David Peat Letter

Merriam-Webster's defines infinite as: "subject to no limitation or external determination." Notfinity provides a basis for an explanation of the Universe's structure and extent. The Universe cannot be a thing, because by definition, things have boundaries. "Universe" refers to an abstraction for matter in motion; what exists and occurs without a beginning, or end. Some support for this assumption of notfinity lies in the fact that there is no evidence to contradict this view.

Matter, which is an abstraction, contains an endless series of smaller microcosms within microcosms ad infinitum. Not only are there microcosms smaller than any object that can ever be discovered, this pattern of microcosms sizes applies also to increasingly bigger structures. Additionally, there is no limit to the number of particles. Thus, this model excludes the concept of a beginning whether through a creation event, such as the Big Bang, or any form of cosmogony.

This pattern never ends, whether getting smaller or bigger. Additionally, there is no limit to the number of particles in this boundless universe. Thus, this model excludes the concept of a beginning of the universe, whether through a creation event, such as the Big Bang, or any other hypothetical event in which the Universe was born.

The idea that the all microcosm-in-motion will eventually disappear as the result of a "Big Freeze," "Big Crunch," "Big Rip" or "Big Change," (in which atoms disappear) are all impossibilities. Because "Universe" represents an abstraction, rather than an actual thing, any microcosm is an instantiation of this, just as any microcosm instantiates the abstraction of matter-in-motion.

David Bohm's conception of infinity included the idea that even the theories we create to describe reality will require endless development to model the new levels of physics being uncovered without end. Physicist

F. David Peat commented on this idea in an August 3, 2014, letter to me.

Dear George,

Thanks for your reply. I remember Bohm telling me about his discussions with Einstein - Bohm has a room in the house next door to Einstein's. Einstein told him that eventually physics would get to the ultimate theory of the Universe. Bohm disagreed he said that for over two hundred years quantum theory had been hidden from the classical world. Below quantum theory would be another level, and below that yet another. These series of deeper and deeper levels would go on forever.

I'm glad that there has been a renewed interest in Bohm's ideas with several international conferences in the last five or six years.

Regards
David Peat

Notfinity also applies to time. When defined in its common, although erroneous way, as something that flows, the noun "time," neither exists nor occurs, because the concept does not represent reality. However, if we convert it to a verb, it serves as a good reference to what occurs, which is the motion of things. Because motion cannot be separated from matter, despite what orthodox physicists accept, we need to understand that "time," when properly defined, refers to matter's motion. Because it makes no sense to say that motion is moving, it is incorrect to say that "time" is flowing or moving. In fact, things are in motion. Moving or timing is not finite, because matter in the form of microcosms limit.

Meaning of Time as a verb

Since ancient Greek times, philosophers, such as Aristotle, have recognized that space and time are abstractions. These thinkers were aware that empty space is impossible. They used it when considering how objects arrange within the Universe. Aristotle and others of his era maintained that time is a measurement of motion or the cycles of change.

Although I also understand that the measurement of motion is important, in defining time it is crucial to remember that what the word time refers to is the moving action. Thus, it would be better to use a word other than the noun time when referring to this.

Einstein stated: "Time is what clocks measure." My response to this statement is to ask what do clocks measure? The indisputable answer is the motion of things, such as the Earth's rotation. Atomic clocks also do this in measuring the motion of the microwave signals from changed energy levels of electrons. A scale measures my weight, but I am a real existing body, not a measurement on a scale. Similarly, a clock measures time, which occurs as motion, but it is not the measurement.

Languages have two parts: one involves matter represented by nouns, and the other refers to the motion of matter designated by verbs. Merriam-Webster defines a verb is: "a word (such as jump, think, happen, or exist) that is usually one of the main parts of a sentence and that expresses an action, an occurrence, or a state of being." I propose changing the classification of the word "time" to a verb to express the idea of the action that is happening rather than the confusing concept of time as a thing, which does an action, as in "passing," or "flowing," etc. This would significantly reduce bewilderment about the concept of time In using time as a verb to mean a type of moving will serves this purpose even better than the word "motion," because that is also a noun.

To help illustrate this use of language, I will again use a tiger. We can express an action of a tiger with the statement,: "A tiger runs." To state what the tiger is doing, we can say: "A tiger is running. Similarly, we say: "The Earth rotates and revolves." To express this in such a way as to indicate that this is equivalent to the occurrence of time, we state: "The Earth times." To phrase this so that it refers to what the Earth is doing, we declare: "The earth is timing."

Reclassifying "time" from a noun to a verb will better reflect its meaning of moving and changing positions. This provides a valid abstraction of what is happening. As a verb, time is suitable to represent the concept of being the action of moving that objects do, which is instantiated in positional change. Despite what orthodox physicists and most people mistakenly believe, time is not a thing, and therefore it is not time that moves.

Just as matter cannot exist without being in motion, the occurrence of "moving" requires the existence of microcosms, which are all included in the matter category. As "timing" refers to this moving, this implies that "timing" cannot occur without matter; and for matter to exist, "timing" needs to be occurring. The two are completely inseparable. Because it is

Theories Without Contradictions

impossible for matter to not exist; it has no origin and cannot disappear..[11]

Microcosms, which are examples of matter, can never stop moving. This is true because even if a particle's motion appeared to stop at 0°K, which is a theoretically impossible temperature, it is still moving relative to everything else. As motion is everywhere, so is timing, but not as a dimension—as in spacetime—or as a thing that exists, passes by or flows, but rather as the moving done by microcosms.

The assumption of an infinite universe offers the potential to bring huge benefits to physics and other disciplines. One of the most significant advantages is that this makes it possible to have a theory for physics and the Cosmos without the contradictions contained in quantum mechanics, Special Relativity Theory and General Relativity Theory. Deriving new implications from such a theory will continue to occur, which will lead to increasingly better approximations of reality.

Because the "infinite universe" is an abstraction, the word "exists" does not properly apply to Universe. Additionally, the idea that the vastness of space is emptiness arises from confusion, because space is not the absence of matter, but rather it is matter in the form of aether.

Understanding the idea of notfinity and accepting it brings freedom from the belief that a concept or theory can fully represent what is not limited in any sense.

Glenn Borchardt[12] made this comment on his website to those who believe that time is an illusion:

> "This is the definition of `illusion: `a thing that is or is likely to be wrongly perceived or interpreted by the senses. Growing up, I used to do some simple magic tricks that clearly were illusions. Nonetheless, I am under no illusion that a great deal of time (motion) has occurred since. The silver hair tells me as much."

To one of his readers who claimed that time is an illusion and therefore so is cause and effect, Borchardt replied:

> "Of course, anonymous is correct to a certain degree: cause and effect do not exist, they occur, as described by Newton's Second Law of Motion. On the other hand, hardly anyone denies causality for some event. This all goes to show how the radical adherence to

logic based on indeterministic assumptions leads to nonsense. Thenceforth, we can consider the statement "time is an illusion" to be pure indeterminism. In addition, like all indeterminism, we can disregard it as psychobabble handed down to thwart the advance of the scientific worldview."[13]

"'Nadal'" and "'timing'" are not illusions, although time as a dimension is a fantasy. This abstraction cannot be instantiated, because time as matter or as part of imaginary spacetime does not exist or occur in the real world. It is a huge mistake to conflate the measurement of motion, which I refer to as nadal`, with the occurrence of time. This is because time involves the actual movement that is being measured, and nadal is just the measurement of it. Motion as a process does not happen in spacetime, but rather in space, which is a form of matter. The idea of time as a thing is not real and thus does not cause this process of changing position. The process, which is a moving, occurs as timing.

Redefining time to refer to moving is more accurate, but requires that one completely purge one's thinking of any association with the traditional meaning for the word time. This is not easy to do, but we can facilitate this happening with enough usage of the word time as a verb.

Presently we use time as a noun. Merriam-Webster defines time as "the thing that is measured as seconds, minutes, hours, days, years, etc."[14] However, change in position of objects is not a thing. The category motion contains all words referring to change in position. Merriam-Webster defines "motion" as: "the action or process of moving or of changing place or position; movement."[15] The dictionary designates motion as a noun, but a noun is not the most appropriate word to express the concept of time as the action occurring. Therefore, using time as a verb can solve this problem, provided we terminate its present meaning.

Glenn Borchardt's phrase "time is motion" expresses exactly the same idea, but because time is a verb in my use of the word, and motion is a noun, I prefer to use the verb "move" to be the same as time.

It is important to understand that moving is a process in which objects are changing their place. Their new position demonstrates the change they have undergone. Thus, the process can be instantiated by the new location of any particular microcosm in space in relation to other microcosms. As particles or objects are never at rest because each relates to the motion of all other particles, there is no moment in which there is no motion.

Theories Without Contradictions

Knowing that "time" refers to "moving," facilitates the realization that time is an abstraction, because moving is. By measuring the motion of one microcosm's change in position relative to another microcosm, it is possible to ascertain a specific time, with the Earth's rotation being the standard since timekeeping began.

The phrase "time equals moving" does not mean that it equates to the definition for time in the spacetime concept, nor the one represented in the flowing time paradigm. To use either of those definitions for time and then think that they refer to the same occurrence as motion is absurd. To understand the meaning of "time is a word for moving," it is necessary to define moving, and then use that definition for time. The generally accepted definition in physics for moving is: changing in position of an object with respect to time. The last four words of the definition are confusing. To define motion accurately, I suggest replacing "with respect to time" with the words "with respect to other objects."

Using this definition, one can then understand that "timing" is a changing in position of an object in relation to other objects. That defines specific time. Universal time is the movement of each particular thing in relation to the position of all other things. Because of the notfinity of things involved, there is no possibility of measuring this movement. It is important to keep in mind that although we can measure "timing", the measurement is not the "timing" occurrence, but simply represents the position change of objects in relation to other objects.

As discussed, nadal means "a measurement of motion" as indicated by clocks. Nadal does not refer to the actual moving that occurs, just the measuring of it, and specifying it as the positional change of an object between each measurement. It facilitates discerning time, which means moving, from nadal, which measures this. The verb **nadalting** means measuring one microcosm's change in position relative to another one. This enables us to ascertain a specific time, which since the first timekeeping devices has used the Earth's rotation as the comparison standard. Until I coined the term nadal, there was no word other than time to refer to the measurement of the movement. The tiniest measurement ever made was 12 alto seconds. Compared to a second, it is the same ratio as a second to about 2.642 billion years. In 12 alto seconds, the Earth travels 0.0000336mm around the Sun at 30 kilometers per second (kps). However, Earth's change in position includes its rotation speed, its speed in moving with the Sun through the galaxy at 200 kps, and with the galaxy through space at 300 kps and the movement in the Universe of the Virgo Supercluster. So even in 12 alto seconds the Earth moves an enormous distance. In addition, it is necessary to factor in the effect of everything moving in the Universe on Earth's position.

Until I coined "nadal," there was no word other than time to refer to the measurement of the movement. The tiniest measurement ever made was 12 alto seconds. An alto second is 1×10^{-18} seconds. Twelve alto seconds to a second is equivalent to the ratio of a second to about 2.642 billion years. In 12 alto seconds, the Earth travels 0.0000336mm around the Sun at 30 kilometers per second (kps). However, Earth's change in position includes its rotation speed, its speed in moving with the Sun through the galaxy at 200 kps, and with the galaxy through space at 300 kps and the movement in the Universe of the Virgo Supercluster. So even in 12 alto seconds the Earth moves more than one would think. In addition, it is necessary to factor in the effect of everything moving in the Universe on Earth's position.

"Nadal" refers to a measurement of the actual change. Having such a term as nadal is extremely important in avoiding confusion about time, because physicists and most others consider time to be the measurement rather than the motion. The term nadal will facilitate discerning time as motion from the measurement of it.

Astrophysicist Julian Barbour, author of *The End of Time: The Next revolution in Physics* writes: "Time is nothing but change. Physics must be recast on a new foundation in which change is the measure of time."[16] Barbour relates time and distance by pointing out that to calculate speed requires dividing it by an actual change in the world, rather than by time, a fact essential to being able to understand the meaning of time.[17]

Barbour is right in stating that time is not than being some sort of substance, but he errs in referring to time as the "quantity" of change, which can be measured by a clock. Although agreeing with him that "time" is not a substance, I maintain that neither is change. Thus, it is incorrect to refer to a "quantity" of change, because that word only applies to things.

Barbour sees time as an average of all changes in the Cosmos, and this amount of change between all nows requires calculating the weighted average of all the differences between them. In Barbour's view, there is change, which is measurable.

I contend that there is moving of things and the measurement of that motion (i.e. nadal). The moving occurs, with "change" being the descriptive result of the motion called time that has transpired. Therefore, it is incorrect to use the two terms interchangeably.

When referring to speed, the same understanding must be incorporated. As time is motion, it is incorrect to refer to things moving in regard to time, as in 100 km per hour. It's not that Earth *take*s a day to rotate, but rather a day refers to a specific rotational distance completed by our planet, which is 40 800 km.

Power of Appropriate Abstractions

Using abstractions enable us to represent a real thing, action or process with an idea or mathematical concept. In theoretical physics, they can help to make the Universe more understandable. However, when not used intelligently or correctly in science, we deceive ourselves into believing that imaginary things are real. The resulting theories of relativity, the Copenhagen interpretation in quantum mechanics, and cosmogony theories, such as the religiously inspired Big Bang Theory, are full of contradictions.

To have the best chance of success in formulating theories, one must appreciate the strengths and weaknesses of abstractions in general. Additionally, to proximate reality with a theory, one must be able to distinguish what is representative of a real category, while realizing that every category is an abstraction, and what attempts to portray reality, but only has imaginary referents. In real categories, it is always possible to have an instantiation, whereas in those of pure fantasy, one can never find real world examples of the abstraction that are supposedly representing   reality or some aspect of it. When used properly, carefully and intelligently, abstractions provide powerful tools for explaining how the Universe works. However, they can easily be misused when employed without comprehending their nature. This can easily result in useless theories that have no chance to advance scientific understanding.

Glenn Borchardt uses valid abstractions in all of his books. These include *The Ten Assumptions of Science: Toward a new Scientific World* view *(2004), The Scientific Worldview: Beyond Newton and Einstein* (2007) and one he cowrote with Stephen Puetz, *Universal Cycle Theory: Neomechanics of the Hierarchically Infinite U*niverse (2011).

In Disruptive:*Rewriting the Rules of Physics* (2016) Steven Bryant offers very valuable information on what abstractions involve. He has considerable expertise in applying the related concept of types, which he utilized in creating Modern Mechanics. He states that MMT's goal is to function as a unified theory of motion.

My opinion of Disruptive appears on its back cover: "A profound shift in the way physicists and cosmologists conceptualize how the universe works. ... I am confident that your book will make a significant contribution."[18] Because these theories are internally consistent, without contradictions or paradoxes, they are far superior approximations of reality than relativity theory or quantum mechanics, which suffer from both problems.

Theoretical physicist David Bohm, who created an alternative interpretation to the standard one for QMT, expressed the idea that the best we can ever hope to achieve in science is better and better approximations of the Universe. These two physics innovators have provided a huge leap forward in this regard.

Language that conforms to Reality

Using time as a verb is more appropriate and accurate than using it as a noun. Having the concept of "nadal" as distinct from time is extremely important to facilitate discerning time, which means moving, from the measurement of it. This helps to avoid confusion, because for physicists and most others, the moving has been confused with the measurement. Nadal" refers to the standard correlated with clocks. Because the confusion in most people's way of thinking of time is so deeply ingrained, it is necessary to again emphasize that the actual moving that occurs in the Earth's rotation and orbiting of the Sun is not a measurement. Use of the word nadal will facilitate discerning time, which means moving, from the measurement of it.

David Peat's Letter on Bohm's Rheomode

To encourage focusing and thinking in terms of processes and holistic movements of things, rather than thinking of isolated objects, Bohm created the "rheomode" as a verb-oriented language form.[19] *Rheo* is the Greek word for flow. The Hopi, Ojibway and Blackfoot tribes use this approach to language.[20] This facilitates a more flowing form of communication than object-based languages, such as English, are able to offer. I am convinced that Bohm had the right idea about the importance of appropriate verb-based language in furthering understanding. However, because of our conditioning to think in terms of objects as being separate from their motion, Bohm's rheomode did not have much success on a large scale.[21]

Someone who shares Bohm's view about language use is David Peat, coauthor of *Science, Order, and Creativity: A Dramatic New Look at the Creative Roots of Science and Life* (1987) with David Bohm, in addition to 19 other books. In *Blackfoot Physics: A Journey into the Native American Universe*,[22] he reveals amazing similarities in insights between native traditions and modern science, while also showing significant differences in the Blackfoot language versus Western European languages that emphasize nouns over verbs. In the first chapter, he writes that in the native sciences of Turtle Island, the concept of process and flux is fundamental. Peat points out that the Algonkian tribes, such as the Cheyenne, Cree, Ojibwaj, Mic Maq, and Blackfeet, use a language that is heavily verb-based to represent direct experience. David wrote about this aspect of Blackfoot language in an August 12, 2015 letter to me relating to my interest in developing more accurate and effective language for conveying ideas. In his letter he refers to my blog entry published in the thescientificworldview.blogspot on mind and consciousness"[23]

Hi George

I enjoyed your article but it also connected me with David Bohm who argued that the use of our Indo European languages, such as English, tends to relevate the Newtonian worldview of well-defined objects in interactions via forces and fields. What he felt was closer to our reality was a language that emphasized verbs and process - what they called the Rheomode (flowing mode). He experimented with this and towards the end of his life encountered the Blackfoot, whose language, which is strongly verb based and worldview, stresses eternal flux and change. In this way, a person's name will change as they engage in different parts of their lives. I remember pressing Leroy Little Bear about the name of a deer. He made a sound. "And is that sound the name of a deer, it signifies a deer?"
"Yes"
Then I tried to make the sound. "So that is a deer".

"Yes, but it could also be a dog".

The Blackfoot language would not let him be pinned down. Process and movement came first; it was not something derived from a static thing put into motion.

I think the same with thought- it doesn't begin with movement. It is movement.

David Peat

Peat originated the concept of gentle action, stating: "Since an objective 'problem' no longer lies outside us, in some external and objective domain, what is now required is an action that arises out of the whole of the situation and is not fragmented or separated from it. Such an action need not be violent but could, for example, arise out of a very gentle, but highly intelligent 'steering' of the system, in which each one of us assumes responsibility."[24]

Quantized Space

Although I support the view that there is no ultimate smallest object, I also share the perspective of physicists who theorize about quantized space at the Planck length. At about 1.6×10^{-35} meters it is almost 100 billion, billion times smaller than the nucleus of an atom.[419] This may appear to be a contradiction, but it is not, because I do not think that there is any ultimate tiniest size of space. In notfinity, quantized space at the Planck size is only the smallest amount of space for things at least as large as the Planck length. There is no valid reason to think that there cannot be another level of quantized space smaller than this. Thus, for objects existing at sub-Planck lengths, there is a quantized space, which has another analogous Planck length that may be 100 billion, billion times smaller than a Planck length.

I think that quantized space has infinite levels. One level involves objects larger than the Planck length. The next level applies to objects smaller than this size. We can further subdivide space occupied by these objects until encountering the next analogous Planck length.

I do not accept the relativity conception of space being part of spacetime, with time being a dimension of this abstraction, which I contend is imaginary. What relativity advocates term as space, I consider as aether, which is a form of matter. Quantized space applies to this aether.

How Specific Motion Occurs in Infinite Space

In the model that I find to be the most rational, which is a realistic interpretation of QMT experimental data, motion is discontinuous. Assuming that only particles are real requires that their realistic motion depicted by QMT must be discontinuous.[25]

In an interview in 1987 with David Peat and John Briggs, David Bohm gave support to this idea. After referring to how classical physics depicts reality as consisting of tiny particles that separate the Universe into independent elements he stated: "Now I'm proposing the reverse that the fundamental reality is the enfoldment and unfoldment, and these particles are abstractions from that. We could picture the electron not as a particle that exists continuously but as something coming in and going out and then coming in again. If these various condensations are close together, they approximate a track."[26] In Bohm's view, it is impossible to separate the electron from its ground, which is all of space.[27]

All motion, including the most specific, is an abstraction that refers to what microcosms (objects) are doing, namely changing position in space. When this new position is related to previous positions of the microcosm, it is clear that this change is part of a series of changes that constitute what we think of as motion. From this explanation, one can understand that motion is not continuous.

Dr. Joel de Rodney, molecular biologist, science writer and futurist, shares my conception of motion. He states: "Intelligence can understand movements or flows only as a succession of juxtaposed still positions."[28] Using the same ideas as I have for years, Borchardt explains motion: "Our sensations come in discontinuous pulses. Vision, for example, is a series of photon impacts that impart information about the shapes and qualities of the material structures before us. The sensation of motion develops in the way in which thousands of still frames in a filmstrip produce a motion picture. Thus, we may see matter, but we can only infer motion. Motion cannot be sensed, for it is not a thing. Only things can be sensed."[29]

The word "motion" designates a process that is inherent to and inseparable from each microcosm contained in the matter abstraction, in which specific examples of matter serially occupy new positions. We measure this motion in many ways, depending on which motion is happening. If the motion is a car traveling on the road, then an odometer measures it. If the motion is the Earth's rotation, then we measure it with clocks.

The measurement size of Planck lengths is useful in explaining how it is possible for an ant to walk across a room. One may think that crossing a room would require traversing across all these infinitely divisible lengths, but that is not the case. In an infinite universe in all directions, if the Planck length was not the smallest space existing for objects of that size and larger, then the distance to cross in a room would be infinite because every distance can be divided infinitely.

This concept comes from Zeno's arrow paradox, which supposedly reveals the impossibility of motion of a thing through space. Zeno, in using the example of an arrow shot at a target, states that in any specified timeless moment, the arrow does not move to where it is, because it is already there, nor does it move to where it is not, because that would require time to get there. He is using word "time" incorrectly to mean something that passes. Therefore at every particular instant there is no motion occurring. He concludes that because everything is motionless at every instant, and time is entirely composed of instants, then motion is impossible.

Zeno was attempting to show that our conception of motion representing things moving through space could not be accurate. There must be a different way to explain the phenomenon of motion.

QMT labels this the quantum Zeno Effect (a.k.a. Turing's Paradox) and attempts to explain it by proposing that time is merely a dimension, which things can move through or remain motionless on a timeline. QMT physicists maintain that through constantly observing a particle it will never decay, which means the observation has prevented it from doing anything and thus what they refer to as "time" ceases. Many studies claim to show that measuring particles with increased frequency affects the decay rate and can potentially stop it completely, which physicists maintain is synonymous with the stopping of "time"

Here is an excerpt from Glenn Borchardt's blog of April 19, 2017,[30] in which he explains the fallacy of this QMT idea:

> "One of the more popular aspects of QMT (quantum mechanical theory) is the indeterministic idea that the experimenter's consciousness might have an influence on the

result. Of course, no one can perform an experiment on any microcosm (xyz portion of the universe) without interacting with it in some way. Time is motion, so whenever one reduces temperature (the vibratory motion of baryonic matter), any motions that occur under normal conditions will be slowed. That is what we do to prevent the decay of food when we freeze it—makes no difference if you are watching your freezer at the same time. And it sure has nothing to do with whether you think time is a dimension or not.

If these experiments have any merit at all, it is that what happens to the microcosm is influenced by the macrocosm. Remember that, according to univironmental determinism, all motions of the microcosm are the result of interactions with the macrocosm.

Also remember that QMT began with the Heisenberg Uncertainty Principle and the observation that both the position and the velocity of a particle cannot be determined at the same time. In other words, any measure of either must involve interacting with the particle, and thus changing its position and/or velocity at the same time. However, this has nothing to do with watched pots or particles. The upshot is that you have to have a pretty big head to think that you can influence the universe much simply by watching it."

Without using the time concept of orthodox physics there is an explanation for this paradox that I have considered since 1980. Julian Barbour discusses it in *The End of Time*. He points out that all objects, including us, do not move from one location to another, because the only true things are "complete positional configurations of the universe, unchanging Nows"[31] In his plausible explanation, he contends that the appropriate view on motion is that the Cosmos changes from one complete configuration or arrangement to another one.[32] As I conceive of all microcosms in constant motion, I have no disagreement with that, as it is a plausible explanation for motion.

He theorizes that there are numberless different instants of time or `Nows` existing simultaneously as points in a timeless, unchanging universe that he calls "Platonia."[33] He accounts for history as being the Universe having gone through a particular series of states.[34]

Barbour uses an assumption that in any moment our experience of motion is a transmuted portrayal of part of a now.[35] Based on his hypothesis that all nows, given once and for all, are complete, motionless instants that simply exist, then motion would not require a transformed representation of part of a now, but simply the next sequential now as occurs in the next frame in a film.

Motion, Time and Change

Moving is what objects do. Object A's total motion results from its own movement and other objects moving relative to it. Without motion, there can be no change, and if there is no measurable change with objects then there is no motion. Change is a descriptive word for the result of motion, which occurs by the arrangement of all microcosms assuming new positions in a constantly reconfigured universe.

Consider the theoretically impossible situation of freezing a microcosm to absolute zero. Even if in some magical way the microcosm could still somehow exist without internal motion, including translational and from zero-point energy, it would still have relative motion, because this is determined in relation to the moving microcosms in the rest of the Cosmos. Of course this entire imaginary scenario is absurd because there can be no matter without internal motion.

Every type of motion brings a change in the microcosm's positional arrangement in space in relation to other microcosms. We can measure the amount of this change that results from moving, which is equivalent to timing, in terms of nadal using any form of clock. For this measurement, the traditional standard by which we have compared any microcosms change has been the Earth's present rotational position or its orbital location in space in relation to the Sun, with a previous rotational position or orbital location. Atomic clocks are the most precise means for determining nadal.

The motion called time involves a series of novel arrangements of the Universe, represented by changes in every microcosm. We experience this change as aging, attributed to accumulated damage from the effects of interactions with internal and external microcosms, which have assumed new positions.

References

[1] Nappi, B. in e-mail comment to me.
[2] Coyne and Borchardt, *Matter and Motion are Abstractions.*
[3] Borchardt, Infinite Divisibility of Matter and Space.
[4] Coyne and Borchardt, Matter and Motion are Abstractions .
[5] Borchardt, in e-mail to me December 15 2916.
[6] Weinberg, *Dreams of a Final Theory*, 67.
[7] Bohm and Callaghan, *Delayed Choice Experiments and the Bohm Approach.*
[8] Ib*id.*
[9] Borchardt, *Indeterministic propaganda against reality*
[10] Ibid.
[11] Coyne, Comment in *Why time is not an illusion.*
[12] Borchardt, *Why time is not an illusion .*
[13] Ibid.
[14] Merriam-Webster Definition of T*ime.*
[15] Merriam-Webster Definition of M*otion.*
[16] Barbour, *The End of Time, The Next revolution in Physics*, 2
[17] Ibid., 96.
[18] Coyne, cover of *Disruptive.*
[19] Peat*, Bohm and the Rheomode.*
[20] Ibid.
[21] George Musser, *The Wholeness of Quantum Reality: An Interview with Physicist Basil Hiley.*
[22] Peat, *Blackfoot Physics: A Journey into the Native American Universe.*
[23] Coyne, *Using Mind and Consciousness in Freedom .*
[24] Peat, F. David, Gentle action.
[25] Qi, *Quantum mechanics and discontinuous motion of particles.*
[26] Peat and Briggs, *Interview with David Bohm.*
[27] Peat and Briggs, *Interview with David Bohm.*
[28] Borchardt, *The Scientific World*view,62.

[29] Ibid., 62.
[30] Borchardt, Quantum Mechanics: A Watched Pot (particle) never boils?
[31] Barbour, The End of Time: *The Next revolution in Physics*, 49.
[32] Ibid., *69.*
[33] Ibid., 44.
[34] Ibid,. 69.
[35] Ibid., 45.

Chapter 9

Physics for the 21ˢᵗ Century and Beyond

What is Dark Matter?

BBT posits that invisible and mysterious "dark matter" permeates the Cosmos as the principal form of all matter. This is essential for BBT; otherwise, indications for the Universe's density are 20 times what astrophysicists calculated as the amount of light elements.[1] Although no scientist has reported the detection of even a single particle of dark matter, nevertheless cosmologists estimate that this non-baryonic matter comprises 84 percent of total mass.

Since 1980, physicists have conducted numerous experiments to uncover proof of its existence, but as of today, there has been no reported direct detection of it. Scientists infer its existence from observed effects of gravity. This refers to motions of visible matter due to gravitational lensing, its influence on CMB, and its effects on cosmic large-scale structures. However, discoveries of warm plasma clouds and white dwarf stars in our local galactic group means that sufficient ordinary matter exists to explain all observed gravitational effects attributed to dark matter. [2,3]

X-ray Emission Line not dark matter

Detection of mysterious X-ray emission occurred in 2014 at about 3.5 kiloelectron-volts in stacked X-ray spectra of galaxy clusters originally hypothesized to come from the decay of sterile neutrino dark matter. If that were true, then all massive cosmic objects would show this spectral feature, but Max Planck Institute (MPA) scientists found no evidence in stacked galaxy spectra for the line, strongly suggesting the theory is incorrect.[4]

Alternative Explanations for Dark Matter

Mordehai Milgrom's Modified Newtonian Dynamics (MOND) Theory interprets the circumstantial evidence for matter differently without hypothesizing its existence.[5] I provided information on this in Chapter 4's subsection "Microwave radiation fluctuations data." It is also possible to make a strong argument that the observed effects are due to aether's presence. Ironically, relativity physicists and BBT proponents have derided the possibility of aether as a form of non-baryonic matter, because relativity does not account for it, but once the BBT needed extra mass in the universe, BBT's proponents quickly decided that dark matter was real, but never using the word "aether " in referring to it.

Maxwell's Aether Theory

In theorizing about aether, James C. Maxwell wrote: "In several parts of this treatise an attempt has been made to explain electromagnetic phenomena by means of mechanical action transmitted from one body to another by means of a medium occupying the space between them. The undulatory theory of light also assumes the existence of a medium. We have now to show that the properties of the electromagnetic medium are identical with those of the luminiferous medium.[6]

Duncan Shaw is one of the most creative and perceptive physics theorists writing on plausible and logical solutions to problems that have stymied physicists for the past century. His 2014 paper, *Reconsidering Maxwell's Aether*[7] published in the journal *Physics Essays*, argues in support of Maxwell's 1865, *The Dynamical Theory of the Electromagnetic Field*, where subatomic aether occupies all space and objects. He shows how Maxwell suggested primary cause(s) for phenomena associated with electromagnetism, and how his aether theory lost favor with the advent of SRT and QMT, which have not physically explained electromagnetism and what comprises its fields. Nor have they avoided immediate action at a distance in entanglement models.

Shaw poses many questions: Why is ordinary logic unable to explain QMT? Why do magnetic fields surround wires that have an electric current? How can we explain dark matter, dark energy and polarization? Is the nature of light particulate or undulatory nature? He proposes that

Maxwell's aether may be able to account for the phenomena associated with each of these rather than just describe effects as QMT does.[8]

Shaw catalogs several esteemed scientists' ideas on the aether concept, beginning with Isaac Newton's[9] treatise *Opticks* about an elastic "aether" that pervades the "heavens" and all "bodies," and transmits light and heat with its vibrations. Shaw mentions that Maxwell deduced aether's existence and referred to it consisting of "parts" and "connections," and acting as a medium for electromagnetic phenomena. He names Paul Dirac, winner of the Nobel Prize in physics in 1933, who was convinced that aether had to exist in some form. Shaw refers to Herbert E. Ives, awarded the 1951 Rumford Medal for his work concerning heat and light, who itemized facts supporting aether's existence. He refers to Stephen Wolfram's depiction of space as a "giant network of nodes" or "cells," and to Robert B. Laughlin, 1998 Nobel winner in physics, who conceived a form of "stuff" and "relativistic ether permeates space. We discover that Frank Wilczek, 2004 physics Nobel Laureate, stated that a "grid" fills space. Shaw informs us that Reginald Cahill contended in *Process Physics: From Information Theory to Quantum Space and Matter* that space's substructure is "quantum foam," and he demonstrated how Michelson–Morley's tests and subsequent ones are supportive of aether's existence. Shaw lists several other notable scientists who have been certain of the reality of aether.[10]

We discover that Frank Wilczek, 2004 physics Nobel Laureate, stated that a "Grid" fills space. Shaw informs us that Reginald Cahill contended in *Process Physics: From Information Theory to Quantum Space and Matter* that space's substructure is "quantum foam", and he demonstrated how Michelson–Morley's tests and subsequent ones are supportive of aether's existence. Shaw lists several other notable scientists who have been certain of the reality of aether.[11]

In his paper, Shaw describes aether as tiny, non-rigid subatomic elastic cells with a vibrating capacity, and ability to form attachments with one another. These structurally identical cells are equivalent to the parts and connections of Maxwell's aether. They are able to vibrate in interacting with other aether cells or with baryonic matter, twist and untwist, and shrink or expand because of their elasticity. Electromagnetic phenomena propagate by their ability to function as organized groups through attachment, and change positions in relation to each other by becoming detached [12]

Polarization of Aether

Shaw discusses some of Maxwell's following discoveries: Electromagnetic phenomena must happen in a substance because action-at-a-distance is impossible. Matter in motion, comprising electromagnetic fields, is aether that occupies all space and bodies and can transmit motion[13] and waves. This medium consists of "elastic yielding" parts and connections. It stores energy by elastic "displacement" of its parts, and from the motion of its parts.[14]

He describes Maxwell's idea that electromotive force causes electric current, which is a type of electric displacement. The stopping of electromotive force results in backflow of current because of the elasticity of electric displacement[15] Rotation of aetherial medium causes rotation in a magnetic field[16]

Shaw explains how Maxwell considered that electromotive force: causes polarization of aether, which vanishes when the force ends. In this model, an electric current's momentum is transferable in two directions between the encompassing electromagnetic field, which connects to it, and the current. Relative energy levels in the current and the field sets the direction[17]

Duncan Shaw quotes Maxwell, "Energy in two different forms may exist in the medium, the one form being the actual energy of motion of its parts, and the other being the potential energy stored up in the connections, in virtue of their elasticity."[18] He also provides this Maxwell quote: "Thus, then, we are led to the conception of a complicated mechanism capable of a vast variety of motion, but at the same time so connected that the motion of one part depends, according to definite relations, on the motion of other parts, these motions being communicated by forces arising from the relative displacement of the connected parts, in virtue of their elasticity."[19]

Aether Theory Implications

Shaw writes that wave-particle duality results from the transmission of energy as waves from individual vibrating aether cells to adjacent ones.[20] Elastic aether cells with interactions and collisions spreading out in every direction, explains how electromagnetic waves transmit.[21]

The broad spectrum of frequencies transmitted by EMR develops from the elasticity of aether cells, which permits their vibrating at several frequencies and in many various directions. Thus, aether activated by interactions and collisions of its cells[22] replaces the electromagnetic field concept.

Existence of aether cells eliminates the need for the concept of the photoelectric effect involving production of photoelectrons, which are electrons or other free carriers, when light contacts a material.[23] This effect is supposedly due to energy transfers from light to an electron. Physicists believe that this indicates photons travel from source to destination. For the waves traveling through an aether medium activating aether cells at the destination, Shaw envisages it as a line of dominoes, each knocking down the next in succession until the last one makes contact with what is at the terminus. He reports that the Compton effect, which results in decreased energy of the "photon," gamma ray or X-ray when the recoiling electron absorbs some of that energy, is also explainable in this same way.[24] He points out that Patrick Cornille concluded that this and the photoelectric effect "can be explained from the classical wave point of view."[25]

Shaw describes how aether allows for an explanation of electric current as the outcome of an electromotive force that causes continuous impacts of aether cells in and along a conductor.[26] He quotes the conclusion of the Einstein–Podolsky–Rosen paper on *Can Quantum-Mechanical Description of Physical Reality be Considered Complete*: "We are thus forced to conclude that the quantum-mechanical description of physical reality given by wave functions is not complete."[27]

As Shaw points out, we can show the physical dimension stipulated by the EPR paper: "If aether is postulated to be a real substance and a medium for the transport of probability waves, this provides `(probability) wave functions` with physical reality."[28]

In his paper, Shaw sees magnetism caused by electromotive force affecting aether cells.[29] In explaining polarization, this same force results in the identical individual aether cells assembling collectively in the electromotive force's direction.[30] This is dissimilar to the standard explanation because the aether medium absorbs the polarization rather than the electromagnetic field.

For a magnetic field encompassing a live conductor, Shaw suggests a possible cause. It is aether cells contracting and expanding at right angles to and in the direction of a current and contorting when in collisions with adjacent aether cells.[31]

Shaw accounts for electricity's back surge following power shutoff with the explanation that it is an electromotive force compressing aether cells when power is on. Then when power goes off, ending the electromotive force, the aether cells bounce back into their regular shape thereby returning the energy stored during compression.[32]

The 2.7° K temperature of the CMB is likely the heat generated by the vibrating and interacting motion of aether cells. This may be what is termed "dark energy."[33] Shaw proposes conceiving of dark matter as being aether by assuming that aether permeates space as an unknown form of matter.[34]

Shaw refers to Roberto Monti, who used experimental evidence to conclude that electric conductivity in space is not consistent with Einstein's traveling quanta. However, it is consistent with aether, which is the source of the 2.7°K CMB temperature and comprises the blackbody,[35] which astronomers refer to as dark matter.

As Shaw mentions, the concept of aether explains how the motion of baryonic matter, referred to as kinetic energy, transfers into electricity. This is because when aether contacts anything, that thing transmits kinetic energy to the aether. For example, falling water's motion causes vibrations, a series of aether cells collisions and that energy is then available as electricity.[36]

Aether moving toward the center of an intervening galaxy or cluster of galaxies may explain lensing,[37] which is light bending from a galaxy or galaxy cluster as it goes by the sides of galaxies or clusters, creating up to three images.[38]

Refraction occurs when there is a directional change in waves when changing mediums. Shaw suggests that aether is a much more acceptable medium to transmit electromagnetic waves than is a vacuum.[39] Other wave phenomena are constructive and destructive interference for which aether may be the medium.[40]

Shaw notes that Richard Feynman[41] observed the same mathematics in describing the phenomenon of water waves in one-slit and two-slit experiments, and those used to depict the interference of electrons in one-slit and two-slit experiments.

The quantization phenomenon may represent structurally identical aether cells' discrete interactions with one another, with atoms and with molecules.[42]

Shaw quotes Patrick Cornille on his view that de Broglie waves are "real physical waves" with structured particles causing quantization.[43]

In concluding his paper on Reconsidering Maxwell's Aether, he affirms that physicists do not conceive of the vacuum as being "empty." Instead, scientists think of it as having physical properties. Therefore, the aether that Maxwell visualized may be filling the vacuum.

Modern Mechanics Theory

Not only did Steven Bryant invalidate relativity, he has also created an alternative theory that has a higher accuracy level than it.[44] This is his Modern Mechanics (MMT), originally known as The Model of Complete and Incomplete Coordinate Systems from 2007. Bryant presents much interesting information on MMT in *Disruptive* where he describes it as a three-system unified theory of motion.[45] This rational and logical model uses a one-directional moving inner-system, a back-and-forth moving oscillatory system and a fixed outer reference system. He highlights its strengths and advantages over QMT, SRT, GRT and Newtonian mechanics.

Bryant explains that MMT derives from the bedrock of geometric transformations (defined in Appendix 2). By explicitly using average intercept times and lengths, and employing a third system to describe oscillation, MMT contrasts with Newtonian mechanics.[46] MMT leads to results that are more accurate by not normalizing its equations.[47]

It differs with relativity theory in several significant ways. Significant differences include going beyond just explaining optics and electromagnetic force, and proposing wave mediums that transmit in excess of light speed.[48] All the absurd abstractions that plague the theories of QMT and relativity physics that I have described in this book, such as the mythical zero-mass photon that supposedly exhibits particle-wave features, and the void of imaginary empty spacetime, are not included in MMT. By not using any of relativity's equations or assumptions,[49] the many problems associated with SRT and GRT do not affect MMT.

MMT also benefits from being highly intuitive. It portrays reality without contradictions or so-called paradoxes, something that relativity and QMT have never been able to do. Its benefits place it in an excellent position to be the replacement for relativity, classical, and quantum mechanics.

Similar to the de Broglie-Bohm pilot wave theory, MMT uses interactions between particles and waves to account for the double-slit experiment's observations. MMT explains valleys and peaks forming when waves emitted at varying times from the barrier go along routes of differing lengths to contact at many points on the observation plane, which observers see arriving simultaneously.[50] Unlike QMT, this interpretation looks at the behavior of each one of multiple waves by examining the pattern created on the plane.[51]

The concepts used in MMT are highly consistent. This easily comprehended theory meets the strictest criteria for valid abstractions of a universe without contradictions. Those interested in theoretical physics will find much that is interesting in *Disruptive*.

The Particle Model

David and Bob de Hilster's model, which uses infinity as an assumption, proposes that a solution to particle-wave duality lies in seeing electrons and light as "plural phenomena," not singular.[52] Because of the way they define light, their model does not require a medium, such as aether for its transmission. It treats light as only the "waves of particles" travelling at identical speed to be waves, rather than always caused by collisions as in aether theory.[53] Its creators see light and electricity to be the same "substance." They contend that this theory is better than aether models for several reasons. These include that it can transmit transverse waves, it describes significant phenomena, including laser light, it does not need matter's density to be extremely high, or elasticity (hardness) to be nearly infinite.[54]

References

---

[1] Lerner, *Big Bang Never Happened*.

[2] Ibid.

[3] Scarpa, M*odified Newtonian Dynamics,* an Introductory Review.

[4] Max Planck Institute for Astrophysics. *Is Dark Matter the Source of a Mysterious X-ray Emission Line?*

[5] Scarpa, M*odified Newtonian Dynamics,* an Introductory Review.

[6] Maxwell, *A Treatise on Electricity and Magnetism*, Part IV Chapter XX.

[7] Shaw, *Reconsidering Maxwell's Aether*.

[8] Ibid.

[9] Ibid.
[10] Ibid.
[11] Ibid.
[12] Ibid.
[13] Ibid.
[14] Ibid.
[15] Ibid.
[16] Ibid.
[17] Ibid.
[18] Ibid.
[19] Ibid.
[20] Ibid.
[21] Ibid.
[22] Ibid.
[23] Ibid.
[24] Ibid.
[25] Ibid.
[26] Ibid.
[27] *Ibid.*
[28] Ibid.
[29] Ibid.
[30] Ibid.
[31] Ibid.
[32] Ibid.
[33] Ibid.
[34] Ibid.
[35] Ibid.
[36] Ibid.
[37] Ibid.
[38] Ibid.

[39] Ibid.
[40] Ibid.
[41] Ibid.
[42] Ibid.
[43] Ibid.
[44] Ibid., 251
[45] Ibid., 280
[46] Ibid., 203.
[47] Ibid., 203.
[48] Ibid., 203.
[49] Ibid., 203.
[50] Ibid., 283.
[51] Ibid., 283.
[52] de Hilster, D. Comparing Aether with the Particle Model.
[53] Ibid.
[54] Ibid.

Chapter 10

Infinite Universe Theory (IUT)

Key Concepts

In formulating the Infinite Universe Theory, Borchardt exposed the absurdity of abstractions of motion separated from matter, hypothetical four-dimensional spacetime, treating time as a thing, the Big Bang Theory, and massive bodies generating their own gravitational fields. He also sees no reason to seek an equation to unify relativity with QMT.

Concepts in IUT include an infinite number of submicrocosms in motion existing in every microcosm, which is part of a macrocosm. Each macrocosm is contained within a much larger supermicrocosm, which is one of a limitless number of these.[1] Each type of microcosm deterministically evolves and changes into other forms, with infinite variety produced by univironmental interactions between microcosms, and between microcosms and their macrocosm.[2]

A key concept of IUT is that aether pervades the three-dimensional infinite Universe, in which redshift arises mainly from absorption, not from the expansion of space as inferred by the discredited BBT.[3] IUT considers light as a wave motion, and its bending in proximity to large masses is actually refraction caused by a dense etherosphere rather than curved spacetime.[4] IUT proposes that all representations of matter, which are everywhere, contain space.[5] It contends that a galaxy's age does not relate to its distance from us.[6] In IUT, abstractions of curved space and the idea of massive objects producing gravitational fields that act as an attractive force are invalid.[7]

There is no concept of a void in IUT, because what is termed space is actually aether, which is a very subtle form of matter.[8] The motion of one microcosm pushes other microcosms, thereby affecting their motion and producing the push aspect of gravity.[9]

IUT conforms to "The Ten Assumptions of Science," as described in Chapter 1, with the most important ones applicable to IUT outlined here. One of these is the assumption of infinity, which refers to the microcosmic and macrocosmic, and avoids all the numerous contradictions inherent in the BBT.

The assumption of causality, where the state of all things links to countless causes, is from David Bohm's infinite universal causality, which is integral to IUT. The assumption of conservation is from the first law of thermodynamics, which does not permit the creation or annihilation of matter or its motion, whereas BBT has the Universe popping into existence from a void. The inseparability assumption arises from understanding that matter or the motion of matter accounts for all phenomena. Thus, it is absurd to consider separating them, because motion is simply what matter is doing. An assumption that the idea of finity is impossible with its intrinsic contradictions helps in furthering understanding.

In IUT, there is no allowance for an existence of nonexistence. I find it amazing and amusing that anyone would support the contradictory idea that there could possibly be a condition or state characterized by not existing.

IUT recognizes that the measured 2.7°K of the CMB, which is the lowest temperature in the Cosmos, proves that there is no void, because empty space has no temperature.[10] Although IUT considers it to be heat emanating from the motion of aether, there is nothing in the theory to preclude molecular hydrogen from being a contributor to the CMB. For IUT the Cosmic Microwave Background is most definitely not radiation from a mythical and impossible Big Bang that has been thoroughly falsified and discredited.

IUT contends that space is matter.[11] This view finds support in Bryant's analysis of the Michelson-Morley data showing evidence of entrained aether.

Infinity in IUT applies to volume, matter, its motion, and the moving of matter as time. Application of the theory leads to no hypothetical beginning point for the Universe, as in BBT or any other cosmogony theories. There is no way to carry out scientific study in those areas.

Borchardt cautions that one must be aware that energy is an abstraction and to not de-objectify the described matter and not objectify its motion.[12] As in all matter-motion terms, such as energy, force, and momentum, these neither exist nor occur. IUT emphasizes that concepts of separating energy from matter or converting it to matter are ridiculous.[13] Just as most people incorrectly think that time is a thing, a similar mistake occurs with the matter-motion terms for energy, which do not refer to things, but rather to calculations done to describe matter's motion. As Borchardt points out, the primary reason for the problem of thinking that these terms represent things is the deleterious influence of immaterialism, which leads to ludicrous ideas of energy, momentum and forces existing as things without spatial dimensions [14]

IUT includes the Univironmental Theory of Gravitation (UTG). It considers gravitation to be equivalent to inertia as Einstein contended in his Principle of Equivalence, which left no role for any attractive force in gravitation. In IUT, convergence applies to infinity of increasingly small and large sizes of microcosms.[15] IUT contends that gravitational pressure keeps the submicrocosms contained within microcosms, such as planets, from moving apart and travelling off separately into space.[16]

UTG employs the concept of gravitational shadowing. In referring to light pressure, George Gamow demonstrated the mechanism. In The *The Gravitation of the Universe*, he writes: 'The pressure of light is very weak as far as bodies of normal size are concerned...there also will be an effect of mutual shadow-casting...Each particle will receive fewer impacts by light quanta coming from the direction of the other one than it will from light quanta coming from all other directions.[17] He proceeds to state that because of mutual shadow-casting, two particles will appear to have an attractive force between them as a result of being pushed toward each other.

IUT contends that because there is no empty space, when a single microcosm moves through the infinite Universe, it affects the motion of other microcosms, and it is the motion of these aether particles that causes the occurrence of gravitation.[18] This gravitation model requires an infinite universe.[19] Newton expressed the consequences of applying gravitation equations to a finite universe: "All matter would fall down from the outsides and convene in the middle."[20] Conceiving of an infinite Cosmos without this form of gravitation is impossible because it necessitates massive bodies generating their own gravitational fields, which violates Newton's first law of motion.[21]

IUT Summary and Comparing Assumptions of Two Theories

As the crowning achievement of univironmental determinism, IUT is more rational, logical and scientific than the religiously inspired BBT, which has violations of scientific laws and assumptions of science. Rejecting systems ideas, logical positivism and the idealism of BBT and relativity in models for cosmology, IUT embraces realism and microcosms.

Borchardt refutes as impossibilities, the abstraction of space as a void, and the existence of independent systems. Instead, IUT sees space as a form of subtle matter referred to as aether, and microcosms—which replaces the word "objects"—existing in a dependent way with their macrocosm that cannot exist without the microcosms it contains. To describe the motions of any microcosm in IUT necessitates examining this relationship known as the univironment.

The following material contrasts the essential characteristics of the Big Bang Theory (BBT) with the Infinite Universe Theory (IUT). This is an adaptation of a chart created by Glenn Borchardt.[22]

Big Bang Theory	Infinite Universe Theory
Finity	**Infinity**
Causality:	
finite universal	Infinite universal causality
Creation	
cosmogony event	**Conservation**
Separability	Inseparabilty
motion without objects	matter-motion is indivisible
Space	
represents void :Yes	No. It contains infinite aether
CMB	
Proves BBT	Proves aether
Non-existence:	
possible	Impossible

Mathematics
  Compliance: Full   Partial

Time is material   Time is the motion of matter

Dimensions
  number: 4 to 13   3

String theory possible: Yes No

Light

Aetherosphere:
  Not present anywhere   Yes

Solar light refraction:
  Due to curved spacetime   Due to Etherosphere

This comparison chart shows the many advantages of the Infinite Universe Theory over the Big Bang Theory, especially in the critical assumptions category. IUT does not violate any of the assumptions of science, but the BBT does. Also very significant is that IUT is able to account for the propagation of light waves through an aether medium, whereas the BBT does not explain how light waves transmit in a void.

The next chapter provides a thorough description of Stephen Puetz and Glenn Borchardt's credible and comprehensible univironmental Neomechanical Gravitational Theory.

References

[1] Ibid.
[2] Ibid.
[3] Ibid.
[4] Ibid.
[5] Ibid.

[6] Ibid.
[7] Puetz and Borchardt, *Neomechanical Gravitational Theory,* Proceedings of the NPA, Albuquerque, New Mexico (2012).
[8] Ibid.
[9] Ibid.
[10] Ibid.
[11] Ibid.
[12] Borchardt, *The Physical Meaning of E=mc².*
[13] Ibid.
[14] Borchardt, *thescientificworldview.blogspot.ca/.*
[15] Borchardt. *The Scientific Worldview*, 190.
[16] Ibid., 190.
[17] Borchardt, *The Scientific Worldview*, 191.
[18] Borchardt, *The Infinite Universe Theory.*
[19] Borchardt. *The Scientific Worldview,189.*
[20] Ibid, 189.
[21] Borchardt, *Infinite Universe Theory.*
[22] Ibid.

Chapter 11

Universal Cycle Theory

Matter's Cyclical Movement

In *Universal Cycle Theory: Neomechanics of the Hierarchically Infinite Universe*, Stephen Puetz and Glenn Borchardt propose a model that is Newtonian mechanics adapted to conform to infinity. They found that the abstraction of infinity for matter, motion and time was indispensable in explaining the Cosmos structure and extent and matter's movement in it without the paradoxes or contradictions contained in the orthodox cosmology and physics theories. Examples of some of the enigmatic phenomena for which UCT provides credible explanations include the propagation of light waves, climate cycles, black holes, dark matter, dark energy, molecular and magnetic bonding, cyclical extinctions, and geomagnetic reversals.

Almost all occurrences in UCT are attributable to matter's cyclical movement.[1] It contends that waves are responsible for linear cycles because of oscillating compressions and decompressions, and every vortex creates circular cycles, by causing matter to rotate.[2]

UCT incorporates the authors' Neomechanical Gravitational Theory (NGT), which includes the effects of vortices. A vortex occurs in a whirlpool, where a mass of fluid having a circular or whirling motion, develops a vacuum or cavity at the circle's center, thereby drawing objects affected by its action to this vacuum.

In GRT, masses deform space resulting in other bodies having to follow curves[3] As in GRT, in UCT there is an equivalence of inertial and gravitational mass. Due to the assumption of infinity and absence of emptiness of space, the motions of microcosms affect other microcosms-in-motion producing in gravitation.[4]

The second law of thermodynamics, which simply is that heat flows from a hotter to a colder body, has a complement. Aether pressure results in gravity. It has similarities, but important differences, to air pressure.[5] The pressure is equivalent to inertia. It can be determined with the inverse-square law.[6]

In the infinite Universe of the UCT, there is no concept of a void because space is not emptiness, but rather it is aether, which is a very subtle form of matter. In UCT, gravitation involves the motion of one microcosm pushing other microcosms, thereby affecting their motion and producing the pushes, which results in aetherial pressure.[7]

The Laser Interferometer Gravitational-Wave Observatory (LIGO), which is a grand-scale physics experiment and observatory to find gravitational waves, announced on February 11, 2016 that a detection had occurred on September 14, 2015. Another was declared on February 11 regarding one from December 26, 2015. A third was revealed on June 1, 2017 to have occurred on January 4, 2017.

The detections were consistent with NGT, except NGT rejects Einstein's explanation for gravity as being fluctuations in spacetime's curvature. Although the media claimed that this proved Einstein's relativity theory, in fact the discovery logically supports the theory that there is aether medium, and that gravitation involves a push, rather than a pull.[8]

Attributing these to a warping of space-time, they did not account for how a void can become warped. Borchardt interprets the results in the June 14, 2017 Scientific Worldview: "All that is happening is the vibration of the aether that pervades the entire universe ...._LIGO is simply detecting the result of a cosmic collision, which has converted some microcosmic motion to macrocosmic motion in the aether. It has nothing to do with gravitation other than that the resulting wave motion is being transmitted by the aethereal medium, which is responsible for the local pressure differences that cause gravitation."

The scientists' contention that the waves "compressed and expanded empty space," is absurd because this would require a medium, which the experiment clearly demonstrated must permeate throughout space. In the UCT this is identified as aether, which is involved in gravitational pressure [9]

Neomechanical Gravitational Theory (NGT)

This Borchardt and Puetz's theory suggests gravitation results from the actions of aether particles throughout an infinite Cosmos.[10] Their activity reduces in ordinary matter because of aether complexes that displace them, thereby creating a type of vacuum for aether.[11] Aether particles have the highest activity and compactness in intergalactic space's mislabeled vacuum, and least in areas of the most baryonic matter, analogous to atmospheric pressure, except in reverse.[12]

This gradient shares some features as the atmospheric pressure gradient, but with significant contrasts. Unlike air pressure, aethereal pressure increases further from Earth's center[13] and from baryonic matter in general. NGT contends that the gravitational pressure gradient (GPG) is the direct cause of gravitation, creating a pressure that acts locally on already present aether particles that permeate and encompass matter. It results in objects being pushed toward massive objects, which act as a vacuum for objects.

NGT proposes that heterogeneous aether particles, having dissimilar values in different directions exist everywhere, with their distributions determined by the pressure gradient of massive bodies.[14] Reduced activity of nearby aether, due to absorption and emission of motion creates pressure gradients making universal gravitation a local phenomenon.[15] NGT uses the concept of the vortex to explain gravitation. When rotating quickly, it creates layering that pushes baryonic matter toward its axis.[16] The theory predicts light velocity is a function of the concentration of aether, and correlates with gravitational and galactic redshift measurements.[17]

NGT Emphasizes Assumptions 4, 5, 6, 8, 9 and 10

Descriptions of these assumptions appear in Chapter 1 in the section The Ten Assumptions of Science. Because NGT assumes infinity, there is no ultimate non-divisible particle or ultimate large cosmic body. Newtonian equations are mostly applicable to NGT, but only as approximations.[18] Assumptions of infinity and relativism are most crucial to gravitation's occurrence. Assuming infinitely divisible representations of matter, NGT proposes a limitless number of aether microcosms in an endless series with each comprised of smaller sub aether particles, constantly moving, while conforming to the implied relativism assumption [19]

Differences from Classical Mechanics

The assumption of infinity distinguishes NGT from Newtonian mechanics. NGT has no constraints. Rather than positing a finite object as in classical mechanics, NGT refers to microcosms, which are any areas of the Universe, all of which contain infinite submicrocosms surrounded by a macrocosm containing a never-ending quantity of supermicrocosms.[20] Anywhere in the boundless Cosmos, every microcosm has submicrocosms that can accelerate with every interaction between any two microcosms of any size, in which both microcosms absorb some of the motion internally.[21]

In NGT Puetz and Borchardt explain that in any group of trillions of particles colliding at or faster than the speed of light, one will be quicker and/ or smaller than other particles because of every microcosm's uniqueness. Due to the shadowing effect, those pushed close to a bigger particle receive fewer impacts from the side toward that particle.[22]

The fundamental cause of bonding and gravitation is in these unequal impacts that particles experience on one side or another, resulting in their rotation around another particle.[23] Stability engendered by reduced movement of much bigger microcosms enables bonding because the combining of different-sized microcosms reduces their previous individual activity, which results in a velocity less than that of most free aether particles.[24] Slower aether particles may evade impacts by combining, thereby forming complexed aether, which is baryonic matter.

The interaction between baryonic matter and very active free aether produces gravitation.[25] Gravitation details arise from differences in relative activity. For example, impacts affect the less-massive combinations of particles more than they influence the larger partner, causing it to rotate around the bigger one. This is an event that occurs at all sizes in the Cosmos because moving small microcosms requires less work than moving bigger ones.[26]

Vortex Formation

Using infinity and relativism assumptions in NGT, the occurrence of the process just referred to indicates it occurs at every scale having comparable univironmental conditions.[27] The gravitation process that changes parts of any kind of matter into complexes of closely bonded matter is termed "complexification." This process, which involves aether, produces hydrogen atoms, comprising 75 percent of baryonic matter that interacts with the macrocosm to become enormous clouds with unique velocity, direction, and mass forming into stars and giant gas planets.[28] If these clouds have glancing collisions with other microcosms, the cloud may rotate, which is crucial to atomic evolution.

While rotating, the velocities of microcosms in the cloud do not change, but move around a core forming a free vortex with the inner region's tangential velocities that are originally identical to the outer area's tangential velocities.[29] This results in the center's angular velocity being higher than the perimeter's, but with shearing between regions, they become equal over time, with vortex rotation forming a forced shape, resembling a solid disc.[30]

Changing to a forced vortex from a free one, thereby further slowing other aether complexes, particularly those of the vortex's core areas, facilitates ordinary matter's formation.[31] In UCT, this process represents matter in motion transforming from one kind to another kind in microcosmic and macrocosmic infinity.[32]

Within the macrocosm of the vortex, very rapid active aether particles create disproportionately greater impacts on the microcosm's side directed toward the macrocosm, resulting in higher aethereal pressures as the radii increases. This causes the gravitational effect that pushes together large, compact matter at the vortex core, which because of increased juxtaposition contributes to more complexification.

Although vortex motion is not essential for gravitation, it helps to bring microcosms together into "complexed aether" accumulations, which through shadowing reduces their exposure to impacts of highly active free microcosms in the macrocosm. Vortex motion and dynamics that exists in the electron[33] and the Local Mega-Vortex[34] dominates the Universe's structures and the motions within the infinite Universe.[35]

Image 7. The Whirlpool Galaxy M51 (NGC 5194) and companion galaxy NGC5195

Credit: NASA, ESA, S. Beckwith (STScI and The Hubble Heritage Team STScI/ AURA

Key Features of NGT

The essential feature of NGT is that free aether exists wherever there is no complexed aether, which is baryonic matter. Aether causes gravitation in a mechanical way. Baryonic matter formation and gravitation fundamentals are complementary processes. Baryonic matter, which is low activity aether complexes, forms from very active aether. Aether's compactness, pressure or activity increases with greater distance from massive bodies. The result is a Gravitational Pressure Gradient (GPG) representing the mass density's inverse function. As Shaw pointed out,[36] this GPG idea originated with Newton in 1718.

Newton's GPG concept

The document on the next page attests to the fact that Newton had conceived of the concept of a gravitational pressure gradient. It is from page 325 in Query 21 of the second edition of *Opticks*. Glenn Borchardt reprinted it in the December 7, 2007 issue of thescientificworldviewblospot.ca.

In spite of propagandists of mainstream science insisting that Newton had no knowledge concerning gravitation's physical cause, and that he considered it an "attraction," this statement by Newton document proves that is a falsehood.

The basis for stating that he denied knowing gravitation's cause is his *hypotheses non fingo* essay in the final 1713 General Scholium addendum to his *Principia,* which was first published in 1687. (The full title is *Philosophiae Naturalis Principia Mathematica* or *The Mathematical Principles of Natural Philosophy*). Francis Motte provided this original translation in 1729:

"Hitherto we have explained the phenomena of the heavens and of our sea by the power of gravity, but have not yet assigned the cause of this power...I have not been able to discover the cause of those properties of gravity from phenomena, and I frame no hypotheses [*hypotheses non fingo*]; for whatever is not deduced from the phenomena is to be called an hypothesis; and hypotheses, whether metaphysical or physical, whether of occult qualities or mechanical, have no place in experimental philosophy...To us it is enough that gravity does really exist, and acts according to the laws which we have explained, and abundantly serves to account for all the motions of the celestial bodies, and of our sea."[37]

Image 8 Newton on Gravitational Pressure Gradient

[325]

Qu. 21. Is not this Medium much rarer within the denfe Bodies of the Sun, Stars, Planets and Comets, than in the empty celeftial Spaces between them? And in paffing from them to great diftances, doth it not grow denfer and denfer perpetually, and thereby caufe the gravity of thofe great Bodies towards one another, and of their parts towards the Bodies; every Body endeavouring to go from the denfer parts of the Medium towards the rarer? For if this Medium be rarer within the Sun's Body than at its Surface, and rarer there than at the hundredth part of an Inch from its Body, and rarer there than at the fiftieth part of an Inch from its Body, and rarer there than at the Orb of *Saturn*; I fee no reafon why the Increafe of denfity fhould ftop any where, and not rather be continued through all diftances from the Sun to *Saturn*, and beyond. And though this Increafe of denfity may at great diftances be exceeding flow, yet if the elaftick force of this Medium be exceeding great, it may fuffice to impel Bodies from the denfer parts of the Medium towards the rarer, with all that power which we call Gravity. And that the elaftick force of this Medium is exceeding great, may be gather'd from the fwiftnefs of its Vibrations. Sounds move about 1140 *Englifh* Feet in a fecond Minute of Time, and in feven or eight Minutes of Time they move about one hundred *Englifh* Miles. Light moves from the Sun to us in about feven or eight Minutes of Time, which diftance is about 70000000 *Englifh* Miles, fuppofing the horizontal Parallax of the Sun to

Y 3   be

It is interesting to note differences between that translation and one from 1999:

> "I have not as yet been able to discover the reason for these properties of gravity from phenomena, and I do not feign hypotheses. For whatever is not deduced from the phenomena must be called a hypothesis; and hypotheses, whether metaphysical or physical, or based on occult qualities, or mechanical, have no place in experimental philosophy. In this philosophy particular propositions are inferred from the phenomena, and afterwards rendered general by induction."[38]

Larry Laudan clarified Newton's meaning for the word *hypothesis*. He explained that in Newton's time, this meant "unproven postulates, axioms or first principles of any science."[39] This was a common definition since Euclid and Aristotle, and Newton's usage of the word in the original *Principia* edition was not disparaging.[40]

Newton historian Dr. Ierome Bernard Cohen explains that Newton occasionally used "hypothesis" to mean the same thing as the word axiom,[41] which means a statement that people accept unquestioningly because it is so well established or evident. Using `hypothesis` in this way enabled Newton to substitute the word "rules" for "hypotheses" in the third edition of *Principia*.[42] Thus, it appears that to some extent Newton is saying that he frames hypotheses, although refers to them as "rules."

About the idea of gravitational attraction, implying action-at-a-distance Newton stated that it is "so great an absurdity that I believe no man who has in philosophical matters a competent faculty of thinking can ever fall into it."[43]

In an e-mail to me on June 25, 2016, regarding Newton's statement, Glenn Borchardt stated that it looks like Newton is "rejecting it in 1718 on p. 325. Indeterminists, of course, forget all about the implied 1718 recantation. In any case: `non fingo` makes no sense at all because mechanics was almost all hypotheses."

Based on Newton's comment, it is possible that in presenting the formula for calculating gravitational results involving two bodies Newton was using the word "attracts" to refer to what appears to occur, rather than as a literal explanation.

Another key feature of NGT is that the Gravitational Pressure Gradient around each cosmic object means gravitation is a universal, but local phenomenon.[44] NGT maintains that rotation involving aether complexes facilitates further complexification, and that resulting vortices create layering, thereby enabling solidification and stratification based on mass density.[45] Additionally, in NGT, light velocity is a function of aether density, and aethereal redshift occurs as light waves pass through regions between galaxies in which the density of aether is highest, contributing to galactic redshift.[46] NGT posits that when light waves go away from a massive body, "aethereal redshift" occurs, whereas when light travels toward a massive body, encountering less dense aether, "aethereal blueshift" results.[47]

Light Velocity a Function of Aether Density

Aether particles responsible for gravitation also transmit electricity and light.[48] Velocity of a wave generally increases in denser mediums as is apparent with sound. Whereas sound waves propagate through a medium of molecules, light waves radiate via a medium of aether. Light waves transmit slower in water because water increases the length of the light ray's path, and its baryonic composition dislodges a higher proportion of aether than air can.[49]

Total-Mass Equation

The "total-mass equation" demonstrates that aether exists where baryonic matter is not present,[50] which is an inevitable outcome of the assumptions of interconnection and infinity, and uses the following parameters: [51]

m_m = Measurable Mass within a specified volume, g/cm^3;
m_{im} = Immeasurable Mass within a specified volume, g/cm^3;
M_t = Total-Mass within a specified volume, g/cm.

$$m_t \to \lim(\infty) = m_m = m_{im}$$

The assumptions of interconnection and infinity imply that in addition to measurable components, matter contains immeasurable ones. NGT explains this by postulating that just as in the case with measurable masses, regardless of how miniscule microcosms are, there are always other smaller submicrocosms in the distance between each microcosm, transmitting matter and its motion.[52] NGT assumes that this same pattern of interconnection applies to infinitely tiny subatomic microcosms because the concept of space without matter is a mere idealization that has no representation in reality.[53]

The total-mass concept identifies the microcosms that act as pushers behind gravitational motions of all microcosms, from the subatomic realm to superclusters of galaxies and larger *ad infinitum*.[54] Throughout the infinite hierarchy, microcosms collide continually with other ones, being both the pusher and the pushed, producing gravitation at every level of the infinite hierarchy.[55]

Diverse combinations of matter in gaseous and solid form are all pervasive, with regions where one form of matter is in high concentration having less of the other form. In passing through the Universe, each microcosm pushes other microcosms of equivalent mass.[56] Each one converges with and diverges from other microcosms according to the complementarity assumption of every motion balancing with other motions.[57]

Gravitation is a Local Phenomenon

The local phenomenon of universal gravitation involves constantly changing combinations of endlessly divisible components of matter encompassing and entering every microcosm.[58] The physical cause of gravitation arises from free-floating aether particles pushing on complexed aether, recognized as ordinary matter.[59]

The Moon and other satellites are within a huge vortex directed by an aether medium that curves the surrounding macrocosm, and they remain in orbit because of greater distal aether pressure compared with proximal aether pressure.[60] Distance from the axis of the vortex that it circulates determines the Gravitational Pressure Gradient.[61]Summary of NGT

As Borchardt and Puetz explain in their NGT paper, from sub subaether particles to structures larger than galactic clusters, all microcosms exhibit the property of gravitation.[62] The assumption of relativism indicates that none of these microcosms is identical to any other. Every aether particle's momentum is unique, as is represented by the formula $P=mv$.

The key to NGT is that faster, less massive aether particles push the more massive, slower aether particles together.[63] As these get closer, they provide increased shelter from hits of less massive particles by producing a complex, which is slower and more ponderous than free aether particles.[64]

NGT explains that because aether complexes are heavy and slow, free aether pushes them toward the center of any rotating cloud of these complexes, which leads to the formation of vortices.[65] Increased size develops these complexes into baryonic matter, which becomes more concentrated at the vortex center, pushing free aether particles to the vortex perimeter.[66] This produces layering, with heavier (solids) complexes centrally positioned and less massive ones (gasses) located peripherally.[67] NGT contends that throughout the process, some aether particles converge and others diverge, with increases in a vortex rotation resulting in matter accretion, and decreases in vortex motion, leading to excreting matter.[68] Microcosms form because of convergence, and disassemble through divergence.[69]

Free aether's density depends on an absence of baryonic matter, which Puetz and Borchardt refer to as "complexed aether." Thus, the concentration of this matter in any particular location results in a GPG with its aethereal macrocosm. Because the impacts from aether are greater from above than from below massive bodies, this results in the pushing of aether toward the center of the vortex, such as to the Earth.[70]

Layering produced by vortex rotation always exists, even though it dissipates with cessation of rotation, resulting in a GPG surrounding every object.[71] Because aether infuses baryonic matter, it is present everywhere to some extent, penetrating through and around all larger microcosms, contacting a microcosm at its most compacted areas.[72] This combined with pressure differences of aether, which are proportional to mass and are local, produces gravitation.[73]

References

---

[1] Puetz and Borchardt, *Universal Cycle Theory*, 97-99.
[2] Ibid, 97..
[3] Einstein, *Relativity The Special and the General Theory*, xxii.
[4] Ibid.

[5] Puetz and Borchardt, *Neomechanical Gravitational Theory, Proceedings of the NPA,* Albuquerque, New Mexico (2012).

[6] Borchardt, *Infinite Universe Theory*, 2007 Proceedings of NPA.

[7] Puetz and Borchardt, *Neomechanical Gravitational Theory, Proceedings of the NPA,* Albuquerque, New Mexico (2012).

[8] Borchardt, *Gravitational Attraction is Dead.*

[9] Ibid.

[10] Puetz and Borchardt, *Neomechanical Gravitational Theory, Proceedings of the NPA,* Albuquerque, New Mexico (2012).

[11] Ibid.

[12] Ibid.

[13] Ibid.

[14] Ibid.

[15] Ibid.

[16] Ibid.

[17] Ibid.

[18] Ibid.

[19] Ibid.

[20] Ibid.

[21] Ibid

[22] Ibid.

[23] Ibid.

[24] Ibid.

[25] Ibid.

[26] Ibid.

[27] Ibid.

[28] Ibid.

[29] Ibid.

[30] Ibid.

[31] Ibid.

32 Ibid.
33 Borchardt, *The Physical Meaning of E=mc²*.
34 Puetz and Borchardt, *Universal Cycle Theory*.
35 Puetz and Borchardt, *Neomechanical Gravitation Theory*. Proceedings of the NPA, Albuquerque, New Mexico 2012.
36 Borchardt, *the scientificworldview.blogspot.ca/*
37 Carey, *Hypotheses (Non) Fingo Sir Isaac Newton's most (in)famous remark*.
38 Wikipedia, *Philosophiae Naturalis Principia Mathematica, Genera Scholium. Third Edition.*
39 Carey, Philosophy Now, *Hypotheses (Non) Fingo Sir Isaac Newton's most (in)famous remark.* .
40 Ibid.
41 Ibid.
42 Ibid.
43 Borchardt, *The Scientific Worldview*, 188.
44 Puetz and Borchardt, Albuquerque, New Mexico 2012. *Proceedings of the NPA* 1 *Neomechanical Gravitation Theory.*
45 Ibid.
46 Ibid.
47 Ibid
48 Ibid.
49 Ibid.
50 Ibid.
51 Ibid.
52 Ibid.
53 Ibid.
54 Ibid.
55 Ibid.
56 Ibid.

[57] Ibid.
[58] Ibid.
[59] Ibid.
[60] Ibid.
[61] Ibid.
[62] Ibid.
[63] Borchardt, *thescientific worldview.blogspot.ca / 2011/12/neomechanical- theory –of- gravitation.*
[64] Ibid.
[65] Ibid.
[66] Ibid.
[67] Ibid.
[68] Ibid.
[69] Ibid.
[70] Ibid.
[71] Ibid
[72] Ibid.
[73] Ibid.

Chapter 12

David Bohm's Approach

Pilot Waves

In 1952, David Bohm had become convinced that QMT could be interpreted using concepts from classical mechanics, so he constructed an electron model on that basis. For Bohm, in apparent disagreement with the prevailing understanding of the Heisenberg uncertainty principle, the electron always has definite position and momentum. An invisible pilot wave field, which is detectable only indirectly through the electron's motions, connects to it and guides the electron. The wave can change shape instantly in response to changes anywhere in the world and communicates this to the electron, resulting in it changing its position and momentum.

The electron has different characteristics depending on the type of measurements performed. This is because these affect the pilot wave, which influences the electron, in such a way that its intrinsic attributes appear contextual due to the pilot wave, making it instantly responsive to all environmental details. The sort of measurement that the researcher is about to perform is one of those details.

Causality and Chance

Bohm's 1957 groundbreaking book *Causality and Chance in Modern Physics*[1] takes an organic view of quantum phenomena, rather than the mechanical one of QMT.[2]

Bohm proposes reality is fundamentally objective and involves an infinite amount of matter in the process of becoming.[3] Because it is

David Bohm's Approach

possible to define its characteristics without reference to anything external, and its existence is not dependent on anything else, Bohm refers to it as absolute.[4] However, in defining nature concretely, he points out that we must consider the relationships of things by which we can make approximate analysis.[5] Bohm maintains that through studying that which is relative in all its limitless diversity and abundance science gets closer to understanding the absolute aspect of the universe.[6]

In describing reality, he opposes the theory that it is based on quantum units of matter that exist autonomously. Bohm's depiction moved the emphasis from material objects to a focus on processes relating to other processes. In this approach, physicists consider particles as semi-independent, apparently local constant aspects of this underlying process, without which the so-called particle and its properties cannot exist.[7] Bohm's model uses all of the ten assumptions of science.

In Bohm's enhanced Louis de Broglie pilot wave theory, it is not appropriate to claim that a wave is guiding a particle, because that involves an incorrect conception of the particle and wave existing independently, which is definitely not the situation.[8]

He provided a causal interpretation of QMT. This made his approach profoundly different from the Copenhagen interpretation, which at the time was the one most commonly held by physicists. Bohm sees mechanistic philosophy in physics as attempting to reduce the Universe to interactions between fundamental entities that have fixed qualities.

Support for Bohm's unorthodox QMT interpretation came in a June 2, 2011, article in *Science*.[9] It describes an experiment where Aephraim Steinberg of the University of Toronto's Centre for Quantum Information and Quantum Control, along with an international research team, used a state-of-the-art measurement technique for the double-slit interferometer experiment.[10] That historic experiment supposedly proved the impossibility of observing a particle pass through a single slit without eliminating the interference effect, and thus one must decide which phenomenon to find. They reconstructed complete paths describing how photons move through the two slits and form an interference pattern. Their technique used a new weak measurement theory developed by Howard Wiseman, who suggested the possibility of determining a photon's direction, based on its observed location. Combining its directional data at many points, enabled an observationally based description of its full flow pattern, including subensembles of quantum particles, as seen on a screen. Steinberg stated that the measured trajectories were consistent with Bohm's realistic interpretation of QMT.[11]

David Bohm contends that extrapolating from successes in science in finding specific mechanistic relationships is unwarranted. As an alternative to these relationships, he proposes the reciprocal relationship between an entity and the wider context it exists in that determines what it is.

Because Paul-Henri Thiry (Baron d'Holbach) was not one to accept arguments independent of experience as constituting evidence, he would have supported Bohm. For Thiry, the Universe is simply matter in motion, determined by unchanging and implacable natural laws of cause and effect. He wrote that there is "no necessity to have recourse to supernatural powers to account for the formation of things."[12]

Bohm's idea of reciprocal relationship has aspects of Borchardt's univironmental determinism, which refers to the microcosm-macrocosm environment. Borchardt defines this as: "That combination of the matter in motion within the microcosm and the matter in motion in the macrocosm that is responsible for the motion of the microcosm."[13]

Bohm shared Borchardt's realization that it is impossible to isolate anything, and all microcosms depend on the existence of other microcosms in the macrocosm.[14] Bohm explains that because the features of all things are dependent on relations to other things, full autonomy cannot exist for any specified thing. He points out that the concept of a thing, supposedly separate from the boundless substructure and background, only exists as an abstraction.[15] This statement touches on two assumptions of science: infinity and interconnection.

Before Borchardt and Elizabeth Patelke devised the word "univronment," Bohm was already employing a similar concept in his book. He contends that matter in the "becoming process," which by definition includes all existing things, is the fundamental reality, and exists independently because its characteristics do not depend on anything external. This understanding enables us to see the Universe's unity.[16] Bohm's idea is consistent with three assumptions of science: materialism, inseparability, and interconnection. It also expresses the concept that matter is always in the process of becoming.

The Bohmian concept of how the whole determines the organized way its parts relate with each other, fits well with the assumption of interconnection. Bohm describes the connection between levels of law: the microscopic and macroscopic and the independent condition present

David Bohm's Approach

with regard to them. This reflects the assumption of interconnection.

Bohm also sees interquantic interconnection in the subatomic world. He refers to "unstable regions" among stable oscillations existing where a system transitions between stable states with rapidity, compared with what occurs in the atomic realm. The speed of the changes means they effectively have discontinuity at one level, although maintain continuous at a deeper one.[17]

Bohm uses the assumption of relativism. He explains that although things have similarities, no two things are identical, because to define something requires referring to an infinity of its characteristics, each of which has some amount of autonomy.[18]

In addition to criticizing the deterministic mechanism, Bohm also disagrees with QMT's indeterminism, in which there is no interest in finding causal interpretations of phenomena. For Bohm, any complete theory must realize that all processes include aspects of causality and chance contingencies, and by focusing entirely on one aspect, we ignore aspects of the broader context.

Bohm shows how causality and chance occur in classical physics. This contradicts the traditional, standard view of classical mechanics, which holds that because of a deterministic state transition dynamics, there is no possibility for chances to be present.[19]

Because standard QMT operates simply as a mathematical algorithm that allows one to perform calculations of probabilities occurring, it does not explain causes for quantum events. Bohm's alternative model,[20] , which involves seeing particles, such as electrons, as part of wave fields, provides the identical predictions as orthodox QMT, but with the advantage of not being paradoxical or contradictory.

Bohm fully supports the assumption of causality. He explains how determinism is workable; even though we may not readily uncover a material cause for a specific event or phenomenon, the cause does exist. He also uses the conservation assumption, stating on page 1 that things never appear from nothing without any antecedents. He asserts indisputably that something always precedes all things, which gives rise to other things in a causal chain; a fact that no observation has ever contradicted.[21] Bohm contends that this overall feature of reality is a principle more fundamental than causality, and without it, there would be no way to comprehend nature with rationality.[22]

For Bohm, the whole is the only existing structure; and the existence of all objects is just the whole expressing itself in relatively independent sub-wholes. These ideas show that Bohm uses the assumption of inseparability. In Bohm's view, the abstraction of the "undivided quantum whole" explains how non-local connections can occur.

The Bohmian interpretation subscribes to an assumption of infinity in terms of an unbounded universe. He also sees this as applying to an endless number of levels of reality. Bohm sees nature as having a qualitative infinity. His own QMT interpretation includes the idea of a sub-quantum level of organization.

There is no evidence that Bohm disagrees with the assumptions of uncertainty, and irreversibility.

Bohm posits that in principle, determinism applies to each object, event or process, but the origin of that determination is the undivided whole, which is self-determined.

The Undivided Universe

In their 1993 book, *The Undivided Universe: An ontological interpretation of quantum theory*,[23] David Bohm and Basil Hiley provide a much-altered interpretation of quantum mechanics. The advantages of their approach is that it does not include problematic features of standard QMT—such as wave function collapse, nor its contradictions and paradoxes, such as wave-particle duality—and does not require an observer for the Universe to exist. However, because it accepts the basic premises of QMT, it does not have any role for aether found in the Infinite Universe Theory, Universal Cycle Theory, or Neomechanical Gravitational Theory.

In the book, they describe their worldview as not completely indeterministic or "absolutely deterministic," considering both of these abstractions to be extremes that represent differing views or facets of the total set of appearances.[24] They suggest that either of the views is appropriate in different situations depending on the type of experiments undertaken and the unknown totality; and that "unlimited essence" ultimately lies beyond both views and our limited abstractions from the whole.[25]

Although I fully agree that abstractions cannot truly model the infinite Universe, I think it is impossible to switch worldviews based on circumstances of situation. I am using the term worldview to mean an articulated philosophical system, fundamental mental orientation, knowledge and viewpoint, including natural philosophy; fundamental, normative, and existential postulates; or themes, values, emotions, and ethics. How can unrelated and contradictory deterministic and indeterministic viewpoints both be included in one's worldview? I would find trying to do that to be extremely confusing in learning how the Universe works. Glenn Borchardt expressed the crux of the problem in an e-mail to me on July 4, 2016: "Compromises between determinism and indeterminism are always regressive, even though they may be sincere attempts at reform."

Bohm's interpretation applied to the measuring procedure avoids the paradox of wave function collapse of standard QMT that supposedly occur because of measurements. The Bohmian concept of how the whole determines the organized way its parts relate with each other, fits well with the assumption of interconnection. His theory includes the idea of the undivided whole, and the "holomovement."[26] This is an essential concept in Bohmian physics. It unites the idea that all matter is in a universal flux or becoming process, with the holistic principle of undivided wholeness, which is a dynamic wholeness-in-motion with everything moving together in an interconnected process. In their book, Bohm and Hiley refer to it as the "primary reality" of the implicate order, from which emerges everything in the explicate order, that we can ever possibly encounter in the Universe.[27]

This concept leaves no possibility of dividing a system from the environment, including the observer, who is not separate from the Universe as a whole. The holomovement is indispensable to Bohm's worldview and his QMT interpretation, uniting the ideas that all matter is in a universal flux or becoming process, and the holistic principle of undivided wholeness, which is a dynamic wholeness-in-motion with everything moving together in an interconnected process.

Bohm's Quantum Potential (QP)

In their January 2015 paper, *Bohm's quantum potential as an internal energy*, Glen Dennis, Maurice de Gosson and Basil Hiley[28] explained the role of Bohm's QP in stationary states and demonstrated that we can conceive of it as an internal energy. They contend that in Bohm's theory the classical point-like particle is not an accurate representation, because the so-called particle is actually an extended structure in phase space, and the QP is an internal energy associated with a particular area of that space. Hiley states, "Each point in phase space represents a state of a classical mechanical system."[29] Also, each state corresponds to one unique point in that space.

In the rest of His August 3, 2017 e-mail to me Hiley continues: "In classical mechanics all dynamic variables commute which means that the order you write variables does not matter. To use this space to describe quantum systems, which is what the Bohm approach does, the order in which you write the variables is vital. We say the phase space in non-commutative. This means that if we use the phase space variables to describe a system, it is not uniquely defined by a point, hence the 'blob'. The properties of the individual become defined by the context. Thus Bohr's notion of 'wholeness' is satisfied. Recall Bohr argued that a description of quantum phenomena must include a specification of the whole experimental set up."

In this model, the QP is part of "context-dependent energy redistribution."[30]

Delayed-Choice Experiments and Bohm Approach

In John Wheeler's delayed choice experiment, immediately prior to an electron's reaching region I 2, a researcher can remove or insert a beam splitter BS 2. When the splitter is not in place, the electron follows one of the paths, triggering D 1 half of the time and triggering D 2 the rest of the time. However, with the splitter present, the electron apparently goes along both paths in a wave-like manner, with all particles directed into D 1 in the resulting interference. Wheeler's claim is that the delay in choice for fixing the beam splitter's position makes the electron decide on behaving as a wave or particle well past the first beam splitter BS 1, although prior to reaching I 2.[31]

Wheeler concluded that "no phenomenon is a phenomenon until it is an observed phenomenon...the past has no existence except as it is recorded in the present."[32] In contrast to this, the Bohm interpretation (BI) accounts for a particle's behavior without such an extreme explanation. Hiley and Callaghan[33] provide an example where putting a microwave cavity in one of the interferometer's arms, results in a particle moving down the other arm and exciting it, implying that the particle not travelling through the cavity surrenders energy. Their paper shows this conclusion to be unwarranted, and a correct application of the Bohm interpretation results in a local explanation.[34] They reason that by properly analyzing the quantum potential structure appearing in the BI, it is clear that only the atom passing through the cavity gives up excitation energy to the cavity even in the BI, and thus descriptions involving non-local energy transfer in the experiments are unnecessary.[35]

Using the assumption that there can be precise position and momentum values, the authors examine its consequences. Bohm and Callaghan state that particles' values for complementary dynamical variables are not the eigenvalues (defined in Appendix 2) of the corresponding dynamical operator, implying that measurement alters the complementary variables.[36] Thus, the act of measuring leads to irreducible and fundamental changes to the examined system, resulting from the QP produced by the measuring apparatus.[37]

Hiley and Callaghan argue logically that we can keep the concept of a particle trajectory in a quantum process by assuming exact simultaneous momentum and position for particles. They write that in the Bohm interpretation the trajectories' equation has QP energy applying exclusively in the quantum realm.[38]

Restricting their paper to the non-relativistic realm, the authors consider the case in which atoms; neutrons and electrons are the incident particles travelling on only one of many possible paths. However, the quantum field, which is one of potentialities described by the waveform ψ, satisfies the globally defined Schrödinger equation.[39]

The BI does not have the problem associated with Wheeler's interpretation of the experiment where the researcher can remove or insert the beam splitter BS 2 as soon as the particle passes BS 1 in the interferometer, but prior to reaching BS 2. Unrelated to whether the researcher inserts BS 2, the particle travels in a single channel, but upon entering the area I 2, its course is affected by whether BS 2 is inserted, determining that region's QP, which determines the particle's behavior.[40]

Hiley and Callaghan explain that if BS 2 is in position when the particle arrives at I 2, then QP enables the particle in channel 1 to go through BS 2, but if it is in channel 2, it will be redirected at BS 2, resulting in every particle entering detector D 1.[41] If the beam splitter is absent when a particle reaches I 2, it will be reflected into a perpendicular direction no matter which channel it is in.

The authors conclude that some criticisms of the Bohm interpretation of the experiments are unsubstantiated, and when M.O. Scully labeled the trajectories' properties as "surreal," he was not using BI correctly.[42] They point out that when used properly, it does not require reference to non-locality to explain the particles' behavior in relation to the added cavity.[43] Hiley and Callaghan mention that they do not know why there is a QP or a reason for the form of the guidance condition. They state that the properties they used come from assuming the particle actually has simultaneous momentum and position.[44]

I find it very amusing and incredibly ironic that Scully, a quantum physicist who is able to accept the enormous absurdities in QMT mentioned in this book, would ever use the word "surreal," in connection to another physicist's explanation for an experiment. Apparently, the Wheeler interpretation of phenomena not existing unless observed is not strange or unusual to him. By any rational standard, everything about QMT is extremely surreal.

References

---

[1] Bohm, *Causality and Chance in Modern Physics.*

[2] Ibid.

[3] Ibid, 170.

[4] Ibid., 170

[5] Ibid., 170.

[6] Ibid., 170.

[7] Hiley, *From the Heisenberg Picture to Bohm: a New Perspective.*
   on Active Information and its relation to Shannon Information.

[8] Ibid.

[9] Science, *Observing the Average Trajectories of Single Photons in a Two-Slit Interferometer.*

[10] *Quantum physics first: Researchers observe single photons in two-slit interferometer experiment.*

[11] Ibid.

[12] Wikipedia, *Thiry, Baron d'Holbach.*

[13] Borchardt, The Scientific Worldview, 124.

[14] Ibid., 122.

[15] Bohm, *Causality and chance in modern physics*, 141.

[16] Ibid., 168.

[17] Ibid., 81.

[18] Ibid., 104.

[19] *Chance versus Randomness, Stanford* Encyclopedia of Philosophy.

20 Bohm, *Causality and chance in modern physics*, chapter 4.
21 Ibid., 1.
22 Ibid., 1.
23 Bohm and Hiley. *The Undivided Universe: An ontological interpretation of quantum theory.*
24 Ibid., 324.
25 Ibid ., 324
26 Bohm and Hiley, *The Undivided Universe*, 382
27 Ibid., 382
28 de Gosson, D and Hiley, B. *Bohm's quantum potential as an internal energy.*
29 Hiley, B. in an e-mail to me August 3, 2016.
30 Ibid.
31 Ibid.
32 Bohm and Callaghan, *Approach*.
33 Ibid.
34 Ibid.
35 Ibid.
36 Ibid.
37 Ibid.
38 Ibid.
39 Ibid.
40 Ibid.
41 Ibid.
42 Ibid.
43 Ibid.
44 Ibid.

Chapter 13

Shaw's Aether Gravity Model

Introducing Shaw's Theory

Chapter 11 presented the case for NGT, which I believe provides an excellent explanation for the way gravitation occurs, and remains the best alternative to GRT. Without in any way undermining my appreciation of the value of NGT, I am presenting an additional theory for consideration, which has some elements in common with NGT, particularly regarding the importance of aether in a pushing or pressuring capacity.

Although there are similarities to NGT it has several significant differences which I will examine here. In presenting Duncan Shaw's theory of gravitation, I am discussing this model with the view that considering more than one credible theory does not have to create confusion for the reader, but rather can generate beneficial discussion, which has the potential for greater understanding.

Flowing Aether

Shaw bases his theory of gravitation on flowing aether, but it is unlike the implausible Georges-Louis Le Sage theory that required aether travelling 100,000 times faster than the speed of light. In contrast, Shaw's theory suggests, "the acceleration rate of gravity at the Earth's surface (9.8 m/s^2) provides a rough indication that the speed of aether flowing into the Earth is relatively slow compared to the speed of light."[1]

In 2012 and 2013, Shaw published two papers on an intriguing gravitation theory in the peer-reviewed *Physics Essays*. They offer strong support for the theory that flowing aether is the physical cause of gravity.

The theory contends that subatomic aether, as a gas and a liquid, permeates all cosmic bodies including space. As a liquid, it flows from space into bodies, creating ram pressure on atomic matter. This pressure is the physical cause of gravity. A crucial aspect of his model is outflowing aether returning to space from cosmic bodies, because it is required to replace the inflowing aether reservoir.[2]

Shaw proposes that aether from space flows in bulk similar to a wind. While pervading all cosmic bodies and space itself, it moves continually between the two while pushing anything in front of it. By converging and accelerating into cosmic bodies, it transfers its linear momentum. This transfer, minus the transfer of linear momentum to atomic matter exerted by expelled aether, is gravity. It is the differential between the aether pressure going into objects and the weaker outward flowing aether pressure.[3] Pushing pressure of inflowing aether on baryonic matter far exceeds that of expelled aether, whose minuscule cells easily pass through baryonic matter and incoming aether with virtually no impact effects.[4]

In this constantly repeating cycle, internal spatial motion pushes aether into cosmic bodies, where its energy dissipates and converts into thermal energy.[5] This results in the expulsion of individual aether cells, which have heat from aether impacts.[6] Therefore, a cosmic body's aether density and pressure falls below spatial aether's pressure, causing the higher-pressure aether in space to flow into the lower pressure, lower density aether bodies.[7] Aether cells' mutual collisions drive aether into lower pressure bodies.[8]

Shaw's vacuum cleaner analogy shows how expulsion causes inflow: expelling air creates a partial vacuum that causes a pressure imbalance between the air inside and outside the vacuum. A vacuum pump results in a pressure differential causing higher-pressure external air to surge into a vacuum. The same process is happening where enormous amounts of cosmic high-pressure aether results in penetration into lower pressure bodies.[9]

In a paper awaiting publication in Physics Essays as of May 12, 2017, titled "Aether Model of Gravity," Shaw incorporates black holes into his gravitation theory. He proposes that the powerful gravity associated with black holes results from the vacuum that occurs when matter is compacted to the tiniest size. As he explained to me, when an object, such as a star, enters a black hole, almost all the area that star material had occupied is now almost completely empty; hence, it is a vacuum into which more matter flows readily.[10]

The Venturi effect explains the accelerated aether flows into cosmic bodies.[11] From beyond the solar system, aether on its way to the Sun pushes on any object in its path.[12]

Shaw shows that because of where momentum occurs, the unequal pushing force on baryonic matter does not violate the principle of equality of momentum.[13] Much greater momentum is transmitted to the matter by inflowing compared with outflowing aether. Single outward-bound aether cells keep their momentum until losing it in collisions with the aether in space with which it merges.[14]

Shaw states that aether flows and accelerates into cosmic bodies in accord with the inverse square law that requires gravity's force to be inversely proportional to the square of a radius from the center of a cosmic body. Similar to the Venturi effect in fluid dynamics, where a fluid's velocity has to increase through a narrowing area due to the mass containing principle, incoming aether from vast space entering a restricted area of a cosmic body converges due to force, which causes increased velocity.[15]

Aether Pressure Differential

As Shaw explains, there is greater pressure by incoming aether on baryonic matter versus expelled aether because they are two states of aether. Bodies expel aether comprised of separate cells similar to gas molecules, whereas aether rushing into bodies from space contains cells that collectively bind together as molecules in a liquid would. This aether type regenerates by condensation. Shaw cites a paper by K. Khaidarov,[16] which depicts aether's transforming into the two states as representing "condensation" and "evaporation." Aether in these two states makes up gravitational fields.

This model shows gravity's direction on Earth is straight toward the Sun. It reveals that the perception of an instantaneous force on Earth and action-at-a-distance is an illusion. It involves aether from beyond the solar system reaching Earth, exerting a push on it directly at the Sun, while flowing along that route, therein setting gravity's direction. The absence of a time factor in gravitational affects supports the flowing aether model of gravity.[17]

The linear momentum of the aether flowing toward the Sun causes a side force on planets. It is at 90 degrees of the direction of the flowing aether, which is in the direction of the orbital path. Side force offsets the friction experienced by planets during their orbits through the aether. Therefore, planets can encounter aether in their orbits, but the aether does not slow down the planet, which would result in it crashing into the Sun.[18]

Some variables determining this force are a planet's shape, momentum, rotation and direction related to aether flow; aether's momentum; effects of aether's acceleration; stream lines and convergence; separation of boundary layer; drag; turbulence; and vorticity.[19]

Assuming that aether causes gravity and it pushes spacecraft returning to Earth, then the ship's speed upon entering our gravitation field must be about equal to inflowing aether velocities. This would mean aether is flowing at 11 kps in our atmosphere.[20]

Aether Experimental Evidence

Shaw points to some of the abundant experimental evidence, such as that of Georges Sagnac, in support of the hypotheses that Earth continuously receives flowing aether on every surface.[21] Providing some substantiation of the idea of aether was Sagnac's conclusion regarding his 1913 experiment with an interferometer in uniform rotation, that "in the ambient space, light is propagated with a velocity, V0, independent of the movement as a whole of the luminous source, O, and the optical system. That is a property of space, which experimentally characterizes the luminiferous ether."[22]

Aether cells' velocity is set at departure and they diverge into space with decreasing intensity according to the inverse square of distance from source.[23] Pound, Rebka and Snider[24] found that radiation entering the Earth is at a higher frequency than the same waves departing. From this, Shaw concludes that radiation travels faster toward Earth than away from it. Assuming that this radiation transmits via a medium of aether, one can logically conclude that this greater speed is evidence that aether is flowing into Earth.[25]

Shaw discusses four reasons that expulsion is indispensable to gravitation. It enables heat dispersion into space, thereby preventing the incineration of cosmic bodies. Expulsion of aether from cosmic bodies causes aether from space to flow into a continuous partial vacuum within these bodies. It provides a source for the necessary constant replenishment of collective aether in space, enabling the gravity cycle to continue. In addition, expulsion prevents bodies from becoming constantly larger from additional accumulated mass.

Because aether cells are so incredibly tiny in their distinct state, they have very few collisions with baryonic matter, whereas incoming aether grouped into billions of cells has a far higher collision cross-section than does expelled aether cells. In reference to this, Feynman states: "The effective `size` of a target in a collision we usually describe by a `collision cross section,` the same idea that is used in nuclear physics, or in light-scattering problems."[26] In applying this idea to both states of aether, Shaw concludes that incoming aether's pushing force is far greater than the pushing of outbound aether.[27]

One may understand that Earth as a cosmic body causes aether's flow pattern but wonder why gravity would not be stronger on the side of Earth exposed to aether arriving from deep space. Shaw's 2013 gravity paper answers this with the concept that Earth is likely cocooned from part of the pushing force of aether rushing toward the Sun. This is because Earth's encompassing aether most likely absorbs the force, and is probably absorbing some force from aether that is on Earth's path around the Sun.[28]

In accounting for gravitation, Shaw also posits convergence, which is an interaction of incoming aether with aether heading sunward. It creates turbulence in the cocooned aether[29] in addition to turbulence due to a fluid flowing by Earth. Cocooning explains why experiments since Michelson and Morley's in 1887 did not appear to provide evidence of aether affecting Earth. He explains this with the fact that experimenters conducted all of their aether tests well within the cocooning aether around the planet.[30] With respect to the Michelson–Morley experiment, Bryant points out in *Disruptive* that it failed because they used an incorrect equation in converting into orbital velocities from raw measurements.[31]

Summation and Conclusions

The aether concept of gravity uses processes of known phenomena with each element in the model bearing upon the others and fitting together to describe gravitation in a rational way. Shaw concludes that there is substantial evidence for the proposal that flowing aether from space into cosmic bodies cause gravitation. This inflow explains the instantaneous gravity between the Earth and Sun, and gravitational acceleration. In drawing from the equivalence principle, Shaw states that objects of varying masses in a flowing river move at identical speeds, which supports the contention that gravity is due to flow toward cosmic bodies.[32] The vacuum cleaner analogy is presented to explain gravitation's accelerating aspect.

Water molecules evaporating and rising through convection and diffusion offer an analogy for how aether gets into space. Michelson-Morley experiments' partial results demonstrate aether's cocooning influence, Evidence of energy that propels inflow is seen in the 2.7° K of space and CMB. Aether's evaporation and condensation into liquid and gaseous states accounts for the unidirectional fact of gravity. Cycling explains aether's replenishment in space and removal of heat from cosmic bodies. Differences in pressure between the aether in cosmic bodies and in space accounts for aether's inflow.

On September 7, 2016, Shaw e-mailed me: "You have done a good job of capturing my gravity concept," He informed me that *Physics Essays* accepted his latest paper.[33] In it, he focuses upon outflowing aether, explaining that heat from ram pressure of inflowing aether brings about its vaporization into gaseous aether; which then flows into space primarily via convection propelled by the buoyancy force, and partly through diffusion. [34]On July 20, 2017 Shaw presented his aether model theory at the annual international conference of the Chappell Natural Philosophy Society, which took place at UBC in Vancouver. BC.

References

[1] Shaw, *The Cause of Gravity—a concept.*
[2] Shaw, Outflowing aether.
[3] Shaw, *Flowing Aether: A concept.*
[4] Ibid.
[5] Ibid.
[6] Ibid.
[7] Ibid.

[8] Ibid.
[9] Ibid.
[10] Shaw, Conversation on progress of *Aether Model of Gravity*.paper.
[11] Ibid.
[12] Ibid.
[13] Ibid.
[14] Ibid.
[15] Ibid.
[16] Ibid.
[17] Ibid.
[18] Ibid.
[19] Ibid.
[20] Ibid.
[21] Ibid
[22] Ibid.
[23] Ibid.
[24] Ibid. .
[25] Ibid.
[26] Ibid.
[27] Ibid.
[28] Ibid.
[29] Ibid.
[30] Ibid.
[31] Bryant, *Disruptive*, 35.
[32] Shaw, W. Aether Concept of Gravity.
[33] Shaw, e-mail to me.
[34] Shaw, Physics Essays.

Part Five   **Consciousness Abstraction**

Chapter 14

Functioning Brain is Consciousness

Neurological Consciousness Study

Neurologists have traditionally defined consciousness as a continual process of arousal and awareness. In November 2016, Beth Israel Deaconess Medical Center and Harvard Medical School researchers claim to have found the neural network that generates consciousness.[1]

Neuroscientists have long known that the brainstem, which is deepest and earliest formed part of the brain, functions to regulate arousal and consciousness. As the starting point for the spinal cord, the brainstem controls the sleep-wake cycle, heart function and respiration. Michael D. Fox, MD, Ph. D one of the researchers commented on discovering a connection between the brainstem and cortex stated. "A lot of pieces of evidence all came together to point to this network."[2]

Researchers showed that the tiny region of the brainstem called the rostral dorsolateral pontine tegmentum performs a crucial function in consciousness. Unconscious patients displayed damage to this area. Dr. Fox commented: "When it is damaged, almost every patient became comatose."[3] Using the human connectome map, neuroscientists found two cortical regions connected to this brainstem area, giving them reason to think that a neural network is formed by these regions, resulting in consciousness.[4]

No one has found the terminus in the cortex for these connections, both of which are associated with awareness. One stops at the pregenual anterior cingulate cortex (pACC), while the other ends at the left, ventral, anterior insula (AI).[5] Up until this study, they were not even implicated in a neural network. These three regions were not functioning in an examination of any of the 45 vegetative or comatose individuals in the study.[6]

Process of Inputs/Outputs and Consciousness

The "process of inputs" into the brain refers to receiving and incorporating raw, mentally unfiltered sights, sounds and sensations. In addition to receiving sensory input data, the brain can also be aware of its own outputs, which are thoughts, emotions and behaviors. The brain's outputs, which are the result of brain processes, are referents for the abstraction of consciousness. There can be no awareness without an object or process as the inputs.

However, there is no possibility that the brain can have direct perception of its own programs, because these exist as a pattern of neuronal connections. We cannot perceive the program for driving a car, although one can be aware of the driving behavior that results from this program. Thus, using a definition of "self" as programs for thinking, feeling and behaving in specified ways; it is not possible to know these programs directly. We can only know the thoughts, feelings and actions arising from them.

For the brain to be free of psychological conditioning it has to stop running these programs. Every time one pursues their selfish interest, the programs are operating and being strengthened through repetition. Those three sentences contain everything you need to know about being free of psychological conditioning. By thinking very seriously and deeply about this you will discover something quite amazing.

If we use the definition of "self" as provided by my friend Bruce Nappi, the president of the A3 Society, then knowledge of this "self" is possible. He defines self as "A felt and visualized experience composed of a stream of data from memory that is passed through the same brain areas as the original sensory data flowed. The data includes past and current experiences, and future expectations, in which the role of the person having those experiences prominent."[7] A big difference in these definitions is mine refers to existing connections within matter as programs, whereas Nappi writes about an occurrence of motion.

Whenever I use the word consciousness, I am referring to an abstraction, The most important fact and key to the apparent mystery of consciousness is that it does not exist or occur as a phenomenon per se, but represents abstractions such as thoughts, feelings, memories and other contents of the mind, along with shortly held, manipulated, interpreted items that survive a filtering process applied to sensory input.

It is impossible for the brain to experience the abstraction of "consciousness" per se. It can only perceive those things contained within the category. What is called "consciousness" is an abstraction for processes occurring in the brain that enable mental activities to occur, rather than being a quality or thing in itself.

"Blindsight" may be illustrative of the concept that the brain demonstrates it can be aware of some incoming data without being conscious of this. It may be providing some evidence that sensory information outside of what the brain seemingly is not aware of can affect behavior, which contradicts the theory that only perceptions that we are conscious of can influence us.

In blindsight, when cortically blind people with primary visual cortex damage or lesions on it receive visual stimuli, they do not consciously see anything, yet they can respond to it. With this type of blindness on a single side of their visual field, they demonstrate greater-than-chance accuracy through guessing or a forced-response scenario, without any consciousness of what someone is showing them. This gives some indication that even though one may not see something, nevertheless the information from visual stimuli is present in the brain. However, it is important to know that the majority of these individuals are only partially blinded, sometimes seeing things, and when thoroughly tested some show partial consciousness of the visual stimuli.[8] Thus, the evidence is not definitive.

The processes involved in the occurrences referred to by the consciousness abstraction are similar in some ways to the sculpting process. Just as sculpting involves eliminating what is not needed, consciousness filters out almost all of the billions of units of incoming data, leaving only four items of content in the global workspace (defined in Appendix 2) during any given two-second time frame. However, unlike sculpting, if at any moment the brain finds something in its awareness that it has screened from consciousness has become important in some way, then it can immediately allow it into its workspace.

It is important to realize that abstractions never exist or occur in and of them self. Because "consciousness" is an abstraction, it does not exist or occur per se. What does occur is a specific thought, perception or emotion. An analogy will clarify this. It is impossible to eat the abstraction "food," one can only eat a specific apple or any other item within the food category once it is in one's mouth. (Because "apples" is an abstraction, one cannot eat "apples") Similarly, one cannot perceive the abstraction of "consciousness." Thus, the idea of being "conscious of consciousness" is not possible. One can only be aware of specific representations of it. I provide more clarification later in this chapter in the section "Hard Problem of Consciousness and Its Solution."

Dualistic Confusion

The abstraction of consciousness has caused great confusion for most people, including neuroscientists, psychologists and philosophers. This mystification is the result of believing consciousness exists on its own as a phenomenon or as a subjective, personal entity as 17th century philosopher René Descartes claimed. In his theory of dualism, the body and mind are composed of completely different materials. A more accurate description for this abstraction is to state that it represents particular experiences and the cohesiveness of our perceptions of the world, which enables a unified view and comprehension of our role in it.[9]

This foundation for the dualistic approach is most often associated with Descartes best known for statement *"Cogito ergo sum,"* which translates as: "I think therefore I am." He argued that it is possible to doubt that one's body exists, but not the mind's existence, because only a mind can doubt. From this, he made the deduction that the mind must be independent of and distinct from the body. That is definitely not a reasonable conclusion, because it requires the unsupportable assumption that the mind is a thing, rather than a pattern of connections in the brain.

A problem arises from the Cartesian conceptual separation of body and mind. The mind, inexplicably considered as a conscious entity within the brain or somehow associated with it, supposedly has beliefs, experiences, desires etc., all of which requires a brain within that entity. Does that entity's brain also have a mind, which has a brain *ad infinitum*?

The brilliant logician, philosopher and mathematician Gottfried Wilhelm von Leibnitz reasoned that Descartes' contemplating the separateness of body and mind did not provide any evidence for the assertion. In order for there to be any validity to Descartes' view, there would have to be examples of the brain's activity not changing while altered consciousness occurs. On the contrary, it is indisputable that many drugs, such as alcohol, marijuana, and LSD that change one's brain activity also affect consciousness. I would be amazed to discover even a single case where changes in brain functioning did not affect one's conscious experience. The fact that even very small changes to the type of thoughts occurring register in MRI brain scans indicates that what is termed consciousness is represented by thought, which is a product of brain activity.[10]

In his 1949 book, *The Concept of Mind*, philosopher Gilbert Ryle uses linguistic analysis to challenge Descartes' myth of the separation of matter and mind, dissolving the concept as being a product of confused language in a fundamental "category mistake." Ryle discusses this type of mistake by referring to a child viewing a division of marching soldiers. After the parents point out the squadrons, batteries and battalions, etc., the child inquires about when the division will come. Ryle comments: "The march-past was not a parade of battalions, batteries, squadrons and a division; it was a parade of the battalions, batteries and squadrons of a division."[11] Descartes made a similar error in thinking that the items referred to in the consciousness category are not the result of brain activity, but rather originate with "the ghost in the machine," as Ryle referred to it. Ryle argued that the brain is sufficient to explain the phenomena associated with consciousness.

Confusion over this separation of body and mind results in the illusion that our way of experiencing is so completely unique that it is not related to that of any other species. This arises from emphasizing differences of complexities over the similarities that we share with them. It facilitates the assumption that the nature of our experiences, which represent the consciousness abstraction, is not an evolutionary process. This encourages a belief that these experiences must be the outcome of consciousness existing per se, presumed to be only available to humans, and requiring religious or supernatural explanations for its origin.[12]

A question that has perplexed scientists since consideration of consciousness began, concerns how in Earth's history matter has gone from being inanimate to having consciousness. A big problem with this enquiry is it assumes that animate matter arose directly from the basic properties of matter, rather than matter having always existed in animate forms. In non-animal species, consciousness refers to the ability to have some response to the macrocosm. For human beings, it is an abstraction for brain processes.

Another problem with the question is the assumption that consciousness exists as a phenomenon, rather than as an abstraction for the occurrence of cognitive activities, such as thinking and experiencing. As a consequence of this unsupported premise, it is impossible to provide an adequate answer to a question not arising from facts. An example of an invalid question would be asking a law-abiding individual whether they had stopped robbing banks.

A more productive inquiry is to ask how the brain has come to have thoughts and experiences. Brain researchers have correlated specific physical states in various brain areas with particular psychological states. However, that does not solve this ultimate mystery; it only demonstrates that conscious states represent the brain's physical state. I propose that the emergence of the ability to experience the world is evolutionary, an inevitable result of the amazing level of structure and organization of the brain to the most subtle degree.

The brain's highly abstract level of thought fosters the belief that we can be free of the sensory world that we live in. From this comes the idea that the desires and wishes we have are not an outcome of our biology and experiences in interacting with our environment, but rather inexplicably, come from a non-material entity that is separate from the body. People may claim to have proof of immaterialism, but any apparent evidence is not part of natural science. This is because that discipline concerns itself entirely with the Universe. A material universe is the only one that can exist, if one is using the word "exists" as meaning something that is material.

Two Minds Hypothesis[13]

This theory proposes that dual parallel processes exist in the human brain. Mind is seen as an upper level system that attempts to meet the organism's goals by creating mental representations of the world and acting upon it. In addition to the "old mind" found in other animal species, we have a "new mind," which has a difficult and occasionally conflicting relationship with the original mind.

The old mind focuses on solving adaptive problems. Its processes are inherent, work rapidly, unconsciously, automatically and use associative parallel reasoning. Contrasting with this, processes in the recent mind respond to rationality, are the result of learning and have flexibility. It is slow, conscious, and sequential, and controlled. It employs rule-based and rational reasoning. Development of this model is hindered by its theorists' poor exchange of ideas and insufficient communication about completed work in social psychology and cognitive psychology.[14]

Consciousness equates to Functioning Brain

Consciousness is the ability of our brain to acquire information and all the content contained in that information with the ability to get all that information into and out of memory. Nothing more is needed to be conscious.[15] The concept of the mind is equivalent to a functioning brain, which is why a diseased or damaged brain results in mental dysfunction. Neurologists account for all aspects of the consciousness abstraction as the result of neurotransmitters reaching their receptors and the firing of neurons. This includes qualia, which is the feeling of what is like to be in a certain conscious state. The motion of electrons fully explains the reason we see one color, rather than another.

The brain can know that it is functioning, but is not aware of how it performs those operations. The phrase "functioning brain" is another way of referring to the concept of consciousness. Full consciousness is a fully functioning brain. Partial consciousness is a partially functioning brain.

There are different ways to conceive of the brain. In the most common concept of a single brain in each cranium then there cannot be "consciousness of consciousness," because this would be the same as saying there can be a "functioning brain of a functioning brain." The brain does not have a brain. It is a brain. The brain is part of the body, so it does not "have" a body. Because a functioning brain is consciousness, there is no "I" that is conscious. I am consciousness, just as I am the body. Understanding this eliminates the source of many problems arising from the unfounded belief in dualism.

Triune Brain Theory

As a result of contemporary neuroanatomical studies, several aspects of this "three brain" model were revised since originally formulated by Paul MacLean in the 1960s. Although this model has fallen into disfavor among a majority of comparative neuroscientists because of its oversimplified organizing theme, it is still worth serious consideration.

This theory proposes that in each cranium there are three brains that have made many interconnections enabling them to communicate and thus influence each other. For example, the limbic system has very highly developed pathways from the cortex to between the cortex and limbic system.

The oldest brain is the reptilian brain, which includes the cerebellum and brainstem. It is compulsive, rigid and reliable; necessary features because it controls respiration, digestion, body temperature and other vital functions. With the appearance of mammals came the limbic brain, which includes the amygdala, the hippocampus, and the hypothalamus. Emotions come from this brain as a result of its ability to create memories of unpleasant and pleasant events. This is the source of value judgments. The flexible primate neocortex (a.k.a. isocortex and neopallium) is involved in higher-order brain functions, such as language,[16] sensory perception, cognition, production of motor commands[17] and spatial reasoning. The neocortex resulted in enormous learning abilities in primates. The two cerebral hemispheres enabled human cultures to emerge with the development of abstract thinking, imagination, and human languages.

If we use the triune brain concept, then the functioning reptilian brain equates to consciousness as do each of the other two functioning brains. The limbic brain can be conscious of the reptilian brain. The functioning neocortex (i.e. consciousness) can include an awareness of both the limbic brain and reptilian brain consciousness. This theory seems very plausible to me.

Global Workspace Theory (GWT)

GWT attempts to account qualitatively for a large set of matched contrastive pairs of neural or psychological phenomena representing conscious and unconscious processes.[18] Blindsight and coma are neural examples. Automaticity with practice, selective attention, and subliminal priming are examples of psychological phenomena.

The global workspace, as theorized by Bernard Baars,[19] contains imagery, inner speech and subjectively experienced events held only a couple of seconds, rather than the 10 to 30 seconds duration of working memory. We can visualize it using a "theater of consciousness" metaphor, with attention being a spotlight on a particular place on stage, revealing the actors who represent the contents of consciousness.[20] Only the actors are visible, everyone else involved in producing or watching the play is in the dark; a metaphor for being unconscious.[21] GWT does not include an observer, or occupy one location in the mind, leading Baars to contend that it is completely different from Descartes" dualistic model.

For sensory consciousness, the lit spot on stage probably needs the involvement of the cortex's regions of sensory projection, and in different senses, consciousness could be mutually inhibitory, within about 100-microsecond time cycles.[22] The "internal senses" of conscious inner imagery and speech results from either external or internal activation of the sensory cortex.[23]

As soon as a conscious sensory item is present, dissemination occurs throughout the brain using expert networks such as "self-systems," which perform as contextual data structures, to receive and mold information from the bright spot. Prefrontal cortex areas, and possibly the parietal cortex for visual orientation, are included in this.[24]

Baars and McGovern contend that the main functional role of consciousness is to enable the operation of "theater architecture" in the brain to facilitate the access, integration and coordination of numerous specialized networks that would otherwise work independently.[25] These are the result of permitting a "blackboard" system, which is "a distributed system of intelligent information processors"[26] to work in the brain. This refers to the problem-solving process where a diverse group comprising specialist knowledge sources begin with a specified problem, update a common knowledge base, (the "blackboard") with partial solutions, and finish with a complete solution.

GWT posits that workspace contents go to numerous subconscious cognitive receiving processes, while parallel processes with restricted communication create coalitions that become input to the workspace.[27] The theory holds that broadcast messages can elicit actions anywhere in the brain by receiving processes, and therefore it is easy to use the workspace to perform voluntary actions.[28] Baars and McGovern explain that to disseminate their messages to all processes to gain recruits for their goals requires workspace access, generating competition between allied and individual processes.[29] Baars points out that because conscious events may require a cognitive executive interpreter, and a "self system," the global workspace is not exactly synonymous with conscious experience.

J.W. Dalton criticized the theory because it does not account for what consciousness is, and how mental processes can be conscious, which is the so-called hard problem." This criticism arises from the unsupportable assumption that the brain has awareness of thoughts and other mental processes. The statement demonstrates confusion in language associated with the difference between thoughts and awareness.

The Conscious Access hypothesis implies that "consciousness" enables use of the brain's multiple abilities in six areas:[30]
1. By using selective attention that can result from the contents of consciousness.
2. By being conscious of events, it is possible to have explicit learning, implicit episodic learning, skill acquisition and episodic learning.
3. Having functions of working memory that depend on visual imagery, self-talk, and conscious perceptions.
4. Accessing the executive interpreters, of the frontal cortex requires being conscious.
5. Accessing comprehensive brain sources results from conscious perception.
6. Having control of complex muscle-and-nerve actions and possibly any neurons with similar properties, including individual neurons.

Chunking Process

A "chunk," as described by Tulving and Craik, (2000)[31] is a familiar collection of more elementary units, frequently inter-associated and stored in memory, and when retrieved act as a coherent, integrated group. They usually have a particular significance to the individual using them. Neath and Surprenant (2003)[32] referred to the "chunking process" as single information units combining to form a meaningful whole. This enables each of the four conscious compartments in the global workspace to contain enormous quantities of information, and through familiarity, mnemonic strategies, pattern detecting and associating new data with one's present knowledge we can boost the GW's performance further. Additionally, in setting priorities attention plays a significant part in enhancing workspace memory capacity by molding brain activity based on its relationship to this objective.[33] However, chunking is far more than simply a means to increase one's working memory capacity; it provides the means to transform raw data into meaningful information.[34]

The first of three aspects of the chunking process involves seeking structures from the information in our working memory. Next is the detection within the global workspace of the chunks of patterned information. Finally, interrelated and hierarchical chunks, constructed in our minds as knowledge over a lifetime, informs attention on what to bring into the workspace, thereby increasing opportunities to find something significant or new that may benefit us.[35] This brings greater effectiveness and proficiency of memory and a positive feedback loop that facilitates discovering novel connections.

Question about Consciousness and the Brain

Realizing that consciousness is entirely a process of the brain has led researchers to ask the questions discussed in this section.

1. Is the key to understanding consciousness in the specific manner that neurons communicate with each other?

Neurons communicate with each other by sending and receiving chemical neurotransmitters, which excites receiving neurons, through synaptic connections or directly via electrical synapses. However, the manner of transmitting the signal is probably not particularly important. What is significant for consciousness is the fact that a single neuron can make 10,000 connections to other nerve cells, which enables highly complex networks to form. However, neuroscientists have not yet learned what types of interactions between neurons produce the features of consciousness.

2. Is the occurrence of consciousness dependent on the entire human brain or only on some areas?

There are two approaches that scientists have in response to this question. One method seeks to find the mechanisms and regions in the brain responsible for "producing consciousness." This group of experimentalists and theorists contend that one particular brain activity generates attention and consciousness. Their hypothesized location for this is in the thalamus, which is involved in processing and facilitating information flow between the various areas of the brain, and focusing attention. These features are all necessary for one to be conscious. Neuroscientist Rodolfo Llinás points out that scanning activity of the thalamocortical system interacting with incoming stimuli generates consciousness.[36] Another group of neuroscientists is primarily involved in learning about the basis and science of attention, chunking and working memory, rather than consciousness itself. Because I see consciousness as an abstraction, representing many experiences, rather than something existing in its own right, this approach of seeking the neural formula for experiences appears to be more useful and appropriate. Trying to find out what creates consciousness is analogous to orthodox physicists searching for the ultimate building block of "matter." In both cases, the scientists are failing to realize that a category does not exist or occur, but represents things and events that do.

3. If only specific regions of the brain generate consciousness, is there some unique quality to those areas?

The thalamus, situated in the center of the brain with nerve fibers extending out in every direction to the cerebral cortex, performs functions, such as regulating alertness, "consciousness" and sleep, and relaying motor and sensory signals to the cerebral cortex.[37]

4. Does the brain only produce one form of consciousness that includes all its aspects, or do separate regions of the brain produce different types of consciousness, such as one for language comprehension, one for visual pattern recognition and another for identifying facial features?

The brain cannot give substance to what is only a category for mental processes, thus it does not create one form or multiple forms of consciousness as a "thing." Just because we customarily group many mental activities together, that does not mean that they are the result of a phenomenon termed "consciousness" but rather they are examples of that abstraction. It is more accurate to discuss the phenomena that we attribute to consciousness. Examples are imagination, perception, thought, processing sensory experiences and emotions, deciphering the meaning conveyed in verbal and non-verbal language, forming thoughts about an imaginary mental self that exists separate to the brain, accessing memories, recognizing patterns and many other mental operations. We also need to include some of the abilities that the brain uses to function, such as attention, filtering, and interpreting, chunking, using working memory and the global workspace, etc. The confusion regarding consciousness arises from thinking that these activities and abilities of the brain need to be features of an overall existing consciousness rather than just mental activities and capacities.

5. What is the minimum number of neurons that can produce some degree of consciousness?

This question assumes consciousness is an existing thing. We know that whatever experiences the roundworm, *caenorhabditis elegans* has, it emerges from only 302 neurons and 7,500 synapses, compared with us with 86 billion neurons, of which 21 billion are in the cerebral cortex, and 1.5×10^{14} are synapses. With at 37.2 billion neurons, long-finned pilot whales have 77 percent more than we have. In principle that should enable these whales to have far more complex, detailed and full experiences than us.

Attention and Consciousness

Attention refers to information filtering that determines what incoming data, such as the billions of units received each second by just the eyes, are important enough for additional processing in analysis and boosting, which enables a control and direction to the focusing.[38]

After attention completes filtering the enormous amount of information, the brain can enhance the workspace's four items, which can be tremendously complex, before using and comparing them to make assessments.[39] When the brain does not use the processes grouped in the consciousness category to identify potential threats, it can apply these processes to focus on food, procreation and social ranking.[40]

In *The Ravenous Brain*, Daniel Bor[41] refers to two experiments that demonstrate that the brain is only aware of what it directs its attention on each moment. In comparison to the enormous amount of incoming data, the range of consciousness is very limited. In Daniel Simons and Daniel Levin's study, volunteers help someone use a map to find a building. Two experimenters walk between them with a door that briefly prevents the volunteer from seeing who he or she has been interacting. During this time of obscured sight, a new person with no resemblance in appearance or clothes replaces the one asking for directions. More than 50 percent of the subjects failed to notice that a different individual is now talking to them. This is because they are focusing their attention inward in working out the directions, rather than on the two men, they are talking with.[42]

In another Daniel Simons and Christopher Chabris experiment, subjects watch a video of people playing basketball while attempting to keep a tally of passes. Someone in a gorilla suit walks slowly among the players, pausing to face the camera and beat his chest. Even though this occupied 5 seconds of the video, 56 percent did not notice the gorilla, because their focus was on watching for passes and keeping count of them.[43]

John Ratey contends that our abilities in areas such as having a sense of self, developing mind, comprehending the world and our relationship to physical and social realities depend on attention and consciousness.[44] Llinás has shown that neurons have the ability to decide how to respond to stimuli. Perceptions are continuously occurring, along with constant activity referred to as consciousness, and the brain is always in a ready state, but it only attends to what it considers relevant.[45]

The features of the consciousness concept emerge when information processing reaches a required complexity and quantity, while attention channels and converts a tiny part of the incoming data into our experiences.[46]

In trying to understand the consciousness abstraction, Ratey provides the analogy of members of a symphony orchestra making discordant noise while tuning their instruments before a performance. Once the conductor hits the baton on the podium, the musicians are attentive, and when the downbeat is given, they cooperate to create harmonious music. This is similar to brain cells transmitting and receiving signals continually without order or direction, as occurs in non-REM sleep, until their attention is aroused, enabling them to resume working as a unit to produce the features and processes associated with consciousness.[47]

In evaluating information, the brain uses short-term working memory, analogous to a computer's RAM to conceptualize events happening now, before analyzing and manipulating the construct in a multitude of ways. The brain also needs long-term memory to provide direction to the present and make plans. These two memory systems permit the prioritization of stimuli, so that some inputs reside in the background of consciousness. They only emerge when needed.[48]

Even when doctors supposedly render the brain "unconscious" with anesthesia, some hypnotized patients can recall verbatim the surgeon's comments made during operations, and studies show those patients have faster recovery times when surgeons make statements during operations that this will be the case.[49] These facts contradict the assumption that the brain cannot receive external inputs in that cognitive state, and demonstrates the distinction between awareness of sensory data and consciousness.

"Free Will" Concept

Those who subscribe to the "free will" concept do not understand that "consciousness" is an abstraction. Just because the brain generates the thought that it is free to decide without influence from the environment, and denotes this imagined "freedom" as being a hypothetical "will," does not prove its existence.

According to the research of psychologists, the self is a mere mental construct. What actually occurs is that the brain continually provides a plausible, although usually inaccurate, explanation of our behavior by creating a narrative self. Largely, inaccessible and unconscious conditioned responses determine our choices. Although these can change with conditioning, the motivation to do this arises from conditioning and other external, deterministic factors in the Universe.

The brain is always making choices and taking action, but because it is part of a deterministic universe any "choice" it makes is an outcome of an infinite series of events that have determined its state in the present moment. If the decision-making was by an immaterial being, then determinism would not apply and "free will" would be possible. But, I do not accept immateralism as real.

The idea that we create the "illusion of free will" is clear. The illusion that we are free to make decisions develops from not understanding how the process of decision-making works and its implications. It involves balancing motivations, including so-called drives, which mostly are outside of our awareness. Therefore, every decision results from reasons based on what we think have more benefits than costs, as determined by the brain's particular conditioning. If "choices" do not come about in this way, it implies that the brain is acting randomly rather than rationally, which excludes the possibility of actions resulting from will.

The statement: "But in producing the result, as long as external restrictions are not imposed on what 'choice' we make, we perceive the decision as freely made."[50] However, there are always external variables involved that impact decisions. In addition, the evolution of the brain is an entirely deterministic process. I did not decide what kind of brain to have at conception. Thus choices are constrained by the nature of the brain. They are also affected by factors such as intellectual ability etc. These internal factors are as important as any immediate external constraints in decision-making.

This does not mean that fatalism is valid. I am a proponent of the univrionmental determinism that Borchardt proposes, in which the environment influences us and we affect the environment. I find this very empowering. Enormous "energy" gets misdirected by thinking that a "self" has to be involved in decision-making. In fact the neocortex makes all "decisions." Freedom is about having no illusions and a state of being watchful. It is not about a trying to exercise a hypothetical "free will."

Additionally, I do not subscribe to the idea of a "social free will," because society is an abstraction for the interactions of all individuals. Thus, as individuals comprising society do not possess "free will," then neither does am=any society.

Eighteenth-century French philosopher Paul-Henri Thiry, Baron d'Holbach made the following argument regarding the illusion of free will, circa 1770.

> "The inward persuasion that we are free to do, or not to do a thing, is but a mere illusion. If we trace the true principle of our actions, we shall find that they are always necessary consequences of our volitions and desires, which are never in our power. You think yourself free, because you do what you will; but are you free to will, or not to will; to desire, or not to desire? Are not your volitions and desires necessarily excited by objects or qualities totally independent of you?"[51]

Baron d'Holbach's hard determinism (HD) position states that because we are material, we are subject to the same immutable natural laws that determine all changes. As our actions result from these determining physical causes, they are not free. The illusion of "free will" arises because the complexity of the many sources of our actions renders it impossible to locate the causes of some of our behaviors.

English philosopher G.E. Moore[52] argues against HD based on its claim that the actions in particular circumstances result from the fact that we could never have done any other action. Moore argues that in HD the concept of "could" or "can" do not exist. However, in his view, we can at times do an action other than the one we do, and therefore HD is not possible. He deduces that eliminating HD leaves the alternative of soft determinism, in which many of our actions result from some degree of freedom and some determination by environmental circumstances and context.

Negative Implications for Free Will from Brain Studies

Brain research experiments suggest that our supposed, "free willed" actions originate from processes occurring in the brain before we are consciously aware of them. Studies support the idea that "consciousness" occurs at a late stage of making decisions, discounting any versions of "free will" in which intent occurs at the start of the decision-making process.[53]

In 2007, neuroscientist John-Dylan Haynes placed people into a scanner that surveyed the entire brain. They then viewed a monitor showing a series of random letters. He instructed them to press a button using their left or right index finger when they felt the urge, and to note the letter on the screen when choosing a hand. The experiment used functional MRI to show real-time brain activity as participants decided which hand to use. The researchers commented on the fact that for many years a controversy has brewed over whether the brain determined subjectively free decisions prior to our consciousness of the decisions. In their conclusion, they state: "We found that the outcome of a decision can be encoded in brain activity of prefrontal and parietal cortex up to 10 seconds before it enters awareness. This delay presumably reflects the operation of a network of high-level control areas that begin to prepare an upcoming decision long before it enters awareness."[54] Thus, prior to the volunteers knowing their supposed free choice, the brain was already preparing the action. Haynes, as well as many other researchers, such as Patrick Haggard, contends that consciousness of a decision may not influence our actions; it could be just a biochemical afterthought, which means "free will" is an illusion.

Haynes refined and replicated his results in two studies. One of them confirmed the brain region's roles with more accurate scanning techniques.[55] In the other, Haynes asked volunteers to subtract or add two numbers from a series on a monitor, which involves a more complex intention than deciding on pressing a button. The experimenters found brain activity as much as four seconds prior to the subjects being aware of deciding. Haynes sees this as a realistic model for real-life decisions.

In treating epilepsy, surgeon and neuroscientist Itzhak Fried implanted electrodes in patients' brains.[56] He detected individual neuron activity of specific brain areas 1.5 seconds preceding their conscious button-pressing decision. Within 700 milliseconds of the choice, the researchers predicted the timing of the decision with greater than 80 percent accuracy. Fried stated: "At some point, things that are predetermined are admitted into consciousness. The conscious will might be added on to a decision at a later stage."[57]

Philosopher Al Mele states: "Part of what's driving some of these conclusions is the thought that free will has to be spiritual or involve souls or something. The trouble is, most current philosophers don't think about free will like that. Many are materialists—believing that everything has a physical basis, and decisions and actions come from brain activity."[58] Thus, in Mele's view, scientists are commenting on an idea that philosophers no longer see as being relevant, because most of them accept that we certainly make rational decisions in a deterministic cosmos. Mele believes that the interplay between freedom and determinism is the only thing to discuss or debate.[59] In considering the possibility that one day, researchers could use brain activity to predict our decisions with 100 percent accuracy prior to our being aware of the decisions, Mele thinks that the idea of free will would be threatened by such an occurrence.[60] Neuroscientist and philosopher Adina Roskies maintains that neuroscience studies indicate predictability of actions, without confirming that determinism is a fact.[61]

Philosopher Walter Glannon speculates on the possibility that one day we may determine causal connections between brain mechanisms and behaviors to replace what are now mere correlations.[62] He states: "If that were the case, then it would threaten free will, on any definition by any philosopher. ...The dualist conception of "free will"' is an easy target for neuroscientists to knock down. Neatly dividing mind and brain makes it easier for neuroscientists to drive a wedge between them."[63]

The brain studies reveal it is a physical system subjected to the same causal forces as any other one, with the implication that its' firing neurons in patterns generates the thoughts and behaviors in a deterministic manner.

Many "free will" proponents assume the existence of an entity in the mind, which is presumably immaterial and therefore not subject to the laws of cause and effect. It is termed the "self," which is depicted as being apart from the brain's processes and programs. This imagined type of self supposedly exists apart from the brain's processes and programs. It is conjectured as exercising its will. In refuting the myth of the ghost in the machine and its "free will," one can focus on the reality of the univironment determining the brain's decision-making.

Neuroscientist Michael Gazzaniga[64] comments that scientists think of decision-making as preparatory brain activity involving a series of small steps. He recommends that researchers develop models for parallel processes with constant interactions in a complex network. He thinks we have put too much importance on the moment of becoming aware of a decision.

Consciousness as Byproduct of Entropy

The second law of thermodynamics states that an isolated system's entropy always increases. Although the rate of change differs between microcosms they are all becoming more disorganized. To maximize chances for survival the brain's development has been directed toward maximizing available information, perceiving opportunities and dangers. However this may have resulted in increasing the entropy rate.

Neuroscientists believe they have identified areas of the brain that could be forming a circuit that generates consciousness. This assumes that it exists in and of itself. Their theory is that it arose as a byproduct of the process of entropy that the brain undergoes. To test their theory, they used a kind of probability theory to look at statistical models of neural networks. This same technique is effective in ascertaining the entropy and thermodynamic properties in other systems. Scientists from Paris Descartes University and the University of Toronto performed tests with nine participants that involved exploring the brain's patterns and how it functions. The purpose was to find out how neurons synchronized their firing patterns, and whether these phased in and out in coordinated patterns in reaction to each other.

Researchers compared two sets of data from subjects who were awake and asleep and observed firing patterns among those having an epileptic seizure. They concluded: "Normal wakeful states are characterized by the greatest number of possible configurations of interactions between brain networks, representing highest entropy values."

In this perspective "consciousness results from how many different ways the brain connects certain bundles of neurons to others, rather than connectivity per se. While the brain was subjected to high entropy, consciousness arose as an "emergent property" to maximize information exchange between neurons. This enables a higher survival rate, but led to an increase in the rate of entropy.

The "Observer" Illusion

Confusion arises from the illusion that a psychological observer exists apart from mental processes and is conscious of the world, and itself. The reality is that only processes occur, which enables observation to transpire. This observing occurs under the influence of the brain's own psychological conditioning, which I refer to as the "self." This results in the brain making interpretations, which may have some commonality with the interpretations others will make who have similar backgrounds, but there is uniqueness in what each brain arrives at.

It appears to me that only human beings carry the burden of an imagined "observer." It is more likely that for other species, consciousness represents their thoughts about what they are doing or planning to do, and feelings that arise in their interactions.

Philosophy of Mind and Predictive Processing

Philosopher Thomas Metzinger of the Johannes Gutenberg University of Mainz, Germany, created the Open Mind project. The second volume, *Philosophy and Predictive Processing* contends that the brain constantly makes predictions about what's in our environment and then we perceive the prediction. The most recent research in the philosophy of mind and neuroscience is that the brain is continually producing what Metzinger and Wiese refer to as a "controlled online hallucination."[65] To have greater predictive accuracy the brain modifies its internal world models or signals the body to move to make the outer environment conform to its predictions. This theory unites perception, cognition and behavior into one model.[66]

A crucial aspect of predictive processing is its intent to challenge the intuitive feeling that our perceptions result from the passive reception of sensory information. Predictive processing posits that perception, cognition and behavior arise from the brain's computations, which are both bottom-up and top-down processing. The theory emphasizes the latter, which involves the way one's emotional/cognitive state and knowledge about the world affects what we perceive.[67]

The brain creates models of the body and the world that it uses to form hypotheses about sensations' sources. What becomes a perception of outer reality is simply the hypothesis considered most probable. When a prediction is wrong, the brain changes its models to explain future situations of a similar nature.

Models, such as the internal organs cannot be changed. For example, sustain a tight body temperature range close to 37 degrees C. the brain predicts that skin sensations will match normal body temperature. If they do not then the brain makes the body go towards cooler or hotter environments to enable the predictions to conform to the physiological condition needed.

In conclusion, the brain is always hallucinating: all our perceptions, including those of our self.[68]

Concept of Self-awareness

The commonly used definition of "self-awareness" refers to the concept that it is possible to have knowledge and awareness of one's personality, character, or individuality, and be able to introspect and discern oneself as separate from others and the environment.[69] In this definition, separate does not imply that the assumption of inseparability is incorrect, just that an animal is aware of being an organism.

There are three levels of self-awareness as defined above. The most basic one applies to any conscious animal. It is recognition of having a distinct existence from everything else, which results in behavior that reflects this understanding.

The second level includes identifying with a particular body and mental characteristics, such as personality, etc. The occurrence of this form is evident by being able to recognize a mirror reflection as being one's own.

The concept of dualism is of fundamental importance in the idea of the third level of consciousness. The brain is aware of the products of its processes, such as thoughts, emotions, feelings, memories and other content. Thus, the term "self-awareness" means that one is supposedly able to distinguish the thoughts from the supposed entity that is separate from the brain that produces them. Fifth century Catholic philosopher Augustine of Hippo expressed the idea of this proposed form of self-awareness: "I understand that I understand."[70] This stated thought implies an awareness of an entity that the brain imagines is producing thought. That is impossible, because there is no such entity; there is only the aware, thinking brain.

After going through the process of analyzing information, which is a major function of the brain, it can comprehend what the data means. However, it is not possible for the brain to analyze its own neuronal connections that form the psychological "self," it cannot actually observe how this program runs. The brain can only observe the way this program affects all aspects of life.

Without an individual claiming to understand that he or she understands something, one can correctly state: "I think that I understand," although others may realize that the one expressing such a view does not actually comprehend what he or she thinks is understood.

Logically, because the brain cannot know itself from an external viewpoint, it will never have objectivity in regard to itself. It can only subjectively perceive how the "self" influences thoughts, feelings and actions.

Another factor interfering in clear self-awareness is the effect of strong emotions, such as rage, which reduce the brain's ability to attend to the subtleties and significance of underlying feelings that precipitated the strong reaction. We can speculate about the reasons for our own behavior, but this is not equivalent to being aware of these. In fact, it is not possible to know all the infinite causes of it.

Additionally, the brain may deliberately not allow into consciousness any thoughts and feelings in contradiction to the self-image, from which it derives an illusion of security. To be fully self-aware would require objectivity, which is not possible, as long as the self (all psychological conditioning) is wired into the brain. If the brain could somehow erase all psychological/social conditioning, then there would be no "self" for the brain to be aware of.

To be objective, the brain would have to distance itself from what was producing the thoughts and feelings, which is the brain. Although most people seem to accept the concept that it is possible to be separate from their own personality in order to observe it, the idea is absurd.

Antonio Damasio's Model

Neurobiologist Antonio Damasio[71] uses "consciousness" as synonymous to awareness." His three-layered consciousness theory involves what he calls the emergence and functioning of "protoself," "core consciousness," and "extended consciousness." As long as one realizes that what he is describing are an organism's specific ways of functioning, rather than a phenomenon labeled "consciousness," there is no problem with his description.

In Damasio's theory, a pre-conscious protoself, which even amoebae have, occurs when neural signals map an organism's internal state onto coherent neural arrangements to produce a fundamental awareness. It notes biological changes that affect its homeostasis.[72] Second-order neural patterns emerge, in which emotions function as neural objects, enabling behavioral responses that alert it to what has transpired within its body. This knowledge, in the form of pictures in its brain, is what Damasio refers to as "feeling."

For Damasio, core consciousness is non-verbal and entirely present-focused. It is an organism's recognition of feelings and thoughts as its own.[73] If one takes a different approach than Damasio, without an assumption of consciousness, this recognition and idea of identifying thoughts and feeling as one's own are simply examples of thought arising from brain processes rather than evidence of the effects of consciousness.

Damasio posits that based on messages from the protoself, the organism constantly creates mental images using communications from the protoself's interaction with stimuli. In this way, a fleeting sense of self appears.[74] The organism keeps making pictures that portray the intrinsic qualities of its experience, enabling it to form a relationship with observed external or internal things. Although Damasio considers a "sense of self" to be indicative of core consciousness, my view is that this is accounted for by the brain thinking about the significance of its mental images.

Damaso's final layer is "extended consciousness," which uses conventional memory to move outside the present. A "narrative self" employs high-level thinking to access previous experiences.[75] This stage of self requires time to be forged. Damasio assumes that being able to retrieve memories requires an "autobiographical layer of self." Although thought is influenced by the narrative self, it does not produce thought. The brain is the only source of thought.

Consciousness Does Not Exist or Occur Per Se

For the brain to fully use and benefit from its abilities encompassed in the consciousness abstraction requires understanding that thoughts and perceptions occur, but these do not imply the existence of consciousness as a thing or occurrence. Not understanding this contributes to the illusion that the mental "self" is a living entity. Believing that self is a live entity, leads to many associated problems that go with it.

A brain that is operating with true intelligence has no illusions. It uses sets of skills associated with the consciousness abstraction to help it deal with and relate to the physical environment and social world. By not identifying with these tools, the brain effortlessly sees through all the illusory beliefs associated with the consciousness abstraction. It is able to have an awareness that uses all levels of thinking and perceiving skills without being confused by thinking that these abilities imply the existence of an immaterial "self" that thinks.

Any attempt to be free of unwanted experiences, such as emotional pain is counterproductive and only reinforces identification with the abstractions of the experience. Mental pain and pleasure involves a form of motion within the brain. This motion, like any other type of motion, does not exist, but instead it occurs. A particular movement within the brain results in an interpretation of the sensation or emotion as either painful or pleasurable. These are both categories of interpretations and as such, they are abstractions.

Once the brain makes the interpretation of certain kinds of motion as being a physical or emotional pain, an additional movement in the brain ensues, which it interprets as suffering. However, if the brain reaches, the conclusion based on its interpretation that what it is sensing is pleasurable, it experiences a feeling of enjoyment. Specific types of interpretations arise as the outcome of specific motions in the brain. However, prior to any interpretation, its internal motion is a response to a stimulus. The specific motion within the brain to a particular sensation or stimulus occurs rather than exists.

A common reaction to unhappiness from mental pain, such as anguish, frustration or sorrow, is to seek what one conceives to be its opposite in the form of calmness, contentment and a sense of joy in living. Unfortunately, this search is as futile as it is problematic, because one is seeking what does not exist, which is true of all abstractions. Therefore, pursuing it as if it is something that one can attain inevitably leads to discontent and frustration. Wanting to have a better psychological/ emotional condition indicates dissatisfaction with one's present experience. The abstraction represents an impossible goal, which creates more unhappiness due to its elusiveness. This process becomes a vicious circle.

Many people take the abstraction even further out and project happiness in heaven or some other projected life without any of the conditions or attributes that define life. In Buddhism there is the abstraction called "nirvana," defined as "the final beatitude that transcends suffering, karma, and samsara and is sought especially in Buddhism through the extinction of desire and individual consciousness."[76] An idea that involves a desire to end desire is contradictory. The very fact that one desires anything psychologically (including the freedom from desire) reveals that one does not see a connection between unhappiness and desire. A truly happy person would simply go about meeting their physical needs and not pursue "nirvana" or any other abstraction of happiness. To make a conscious effort to extinguish individual consciousness is contradictory and absurd.

Those who think that they have succeeded in this effort and believe they are enlightened fail to understand that this term is an abstraction, and because all abstractions have no existence or occurrence apart from thought, enlightenment does not exist. Thus to seek it is a delusional process.

This pursuit will inevitably contribute to a strengthening of troubling processes that involve preoccupation with one's mental "self." This only leads to the emotional pain of feeling isolated from others and life. Additionally, overall anger and annoyance will increase, because one will fail in trying to get something that does not exist. Even if one fools oneself into thinking that he or she has attained success, there is no escape from the reality of their actual mental condition; they will just be numbing their emotions, resulting in a condition that is worse than having frustration, because now they are less in touch with who they are in reality.

Functional Explanation of Easy Problem

Rather than accepting the validity of the "consciousness as a thing" abstraction, I refer to the occurrence of experiences associated with a variety of phenomena. Some of these, which are just concerned with mental functions and abilities, are easily explainable with a neurophysiological or cognitive model as being the outcome of neurobiological computation.[77] Some examples are degrees of wakefulness, attentiveness, and the capacity to respond to stimuli. Other examples are the awareness of and ability to communicate thoughts and emotions, the ability to assimilate facts and experiences, and the ability to manage one's behavior.[78]

David Chalmers provides an example of how to explain functions in *Facing Up to the Problem of Consciousness*[79] where he describes control and integration that models how the brain's central processes organize information to facilitate specific behaviors. Explaining a function's performance only requires identifying the mechanism executing the function, either in neural or computational terms.[80] Being appropriately receptive to environmental inputs that enable using that information in relevant behavioral responses designates being awake.[81]

By describing how DNA stores and transmits inheritable traits, we can explain the gene.[82] Brain researchers explain learning by depicting how a neural or computational mechanism enables one to change behavioral capabilities when presented with information and use it to adapt to the environment.[83] They model language, memory, perception and other mental phenomena in a similar fashion.[84]

Hard Problem of Consciousness and its Solution

In 1995, David Chalmers started using the term "hard problem" for the long-standing problem of explaining why a physical state is conscious, including why conscious mental states "light up," and the occurrence of "something it is like" when subjects have conscious experience.[85] In spite of accounting for the conscious brain's structural, functional, and dynamical features, which include how it is assembled, what it does, and how it changes, the question of why it is conscious has persisted, apparently exceeding the explanatory capacity of science.[86]

The hard problem also includes the question of why or how we experience "qualia." Because these refer to subjective experience—such as the taste of a pear, the perceived blueness of the ocean, the pain of a burn—describing them is difficult. The minds of conscious beings supposedly perceive these experiences. Researchers can show links between physical processes between neurons that explain the way qualia form and their location, but do not know why we experience them.[87] The hard problem is about how to explain subjective experience.

This problem does not deal with the brain's cognitive abilities or causal part in generating behavioral performance. Therefore, for conscious experience, even after scientists elucidate the applicable functions, they have difficulty creating a model for it.[88] For example, they can explain many aspects of the brain's function, but not answer why there is experience, rather than none.[89] After light enters a retina, sending electrical signals to the brain, why do perceptual discrimination, internal access, categorization, plus verbal report produce a subjective experience?[90] Neuroscientists and philosophers consider this the primary question in the problem of consciousness.[91]

There has been a lengthy history of difficulty in trying to explain the emergence of subjective states of experience in our mental processing of data from external and internal sources. These include physical and perceptual sensations, images and thoughts generated by the brain, and the experience of emotion. This so-called hard problem of consciousness is actually quite explicable.

In beginning to respond to this apparent problem, it is important that we not allow assumptions of dualism to constrain us, or to accept the question's premise. If one does assume that subjective experience is different and separate from perceptual discrimination, internal access, categorization and report, then all we need to do is examine these processes, and not be concerned with the abstraction of the hard problem concerning consciousness. Here is an analogy that will illustrate the point: Put several raw, chopped vegetables into a bowl, and add dressing or oil. Would it be worthwhile to ask: "After someone combined these vegetables, what did the vegetables then do to produce the result that we label a salad? The answer is that the mixture of vegetables did not do anything to produce a salad, but rather, after mixing the vegetables together, we call it a salad. Asking how the brain produces subjective experiences is similar to asking how does lifting each leg alternately in rapid succession, while thrusting the body forward horizontally produce a phenomenon termed running. The motion of the body is occurring, and "running" refers to that motion. Using the same logic, we can understand that perceptions and awareness refer to the same occurrence. Perceptions do not produce a state of being aware. This term simply describes what transpires when stimuli enter the brain. That state or condition of the brain after the filtering and analytic process takes place on a continuing basis is what the term "consciousness" equates to, which is always subjective due to the presence and influence of the self and mind.

In the process of absorbing and understanding information, the brain continually transmits the data through its entire neuronal structure. The cerebral cortex evaluates the information based on how well it fits with established patterns that comprise the mind. After the brain filters the raw data through a very individualized mind structure, the data takes on a new significance. The brain interprets the output, and a pleasant, unpleasant or neutral experience occurs.

In summary, the subjectivity of the experience arises from the process of the brain interpreting the original information according to its own mental constructs. Thus, the "hard problem of consciousness" resolves itself through understanding how the mind influences what the brain is aware of at any given moment.

Motion is always about what matter is doing. In the case of thinking, the motion is in reference to the activity in the brain. This involves the firing of interconnected neurons in a network enabling them to communicate with each other and with the network.

Conceiving of consciousness as a "thing," rather than an abstraction, makes clarity in this area difficult. The confusion can lead to a belief that an independently existing mental self is the ultimate source of thoughts, and that this entity can exist after the death of the body.

Mental phenomena occur as the result of the motion in the brain, which leads to all the thoughts and experiences that we classify in the category of consciousness. It is not possible to locate any area of the brain that is the focal point for the existence of a hypothetical thing identified as consciousness, because consciousness represents a category of experiences.

The totality of the brain's responses to external and internal data resulting in mental and physical behavior and experiences, including remembrances of experiences, is the reality of what is in the brain. In thinking about our experiences, we come to believe erroneously that there is a thing termed consciousness that has separability from experiences. Although this illusion can seem real, it is impossible, because it would violate the "fourth assumption of science," which is inseparability.

It is not accurate to refer to "I" when discussing consciousness, because all that actually occurs is experiencing. The "I" is just a concept attached to the experience. Thus, the solution to the so-called hard problem of subjectivity in consciousness lies in the particular connections in the brain that comprise the mind. In this model, the word mind signifies the unique neural network that exists in every human brain.

The brain exists as organic matter, and the mind is the pattern of connections in that matter. Although the mind is analogous to the outputs from a computer, it does not actually exist as a thing. The totality of connections forming the mind does not create a physical object, but merely represents a relationship of connections between individual brain cells with each other and the entire structure. The mind is not a living entity within the brain or an "organism within an organism."[92] However, without a mind a brain would be as useless as a computer without the ability to generate outputs. The mind is the neuronal pattern that has formed and continues to develop from the connections between brain cells. That is a very important concept to understand. Unfortunately, very few people ever realize the significance of that statement and its implications about being human.

As is the case with any material structure, for its existence the mind requires internal motion. Moving electrons within brain cells meet this need. The continual process of dendrites reaching out to form new connections provides movement of its components. The mind's contribution to the brain's interaction with input from the environment enables meaningful experiences to occur. Within the framework of the mind, the brain interprets these experiences in particular ways, which we generalize with the abstraction of consciousness.

Many of the connections forming the mind are already present at birth, having become part of the inherited hard wiring of the brain. The rest come about because of interactions between the organism and its environment. Our unique mental framework, including the ego and self-image constructs, determines the particular interpretation and incorporation of our observations.

The structure of the mind determines the way the brain interacts with the world. Every encounter with the external environment, including other people, affects the way the brain forms its mind, which in turn influences the way it meets with and understands the world. This is a continual process, in which the brain is constantly making more connections to this wired mind. These modifications influence the organism's behavior and subsequent approach in encounters with the environment, which leads to additional changes in the construct of the mind.

Neuroscientists maintain that consciousness is a product of the brain's activity, but this abstraction is insufficient, because it implies that it is separate from the brain's functioning. When the neurons are firing in communication with one another and their network, this process involving the entire brain/mind complex generates the great variety of phenomena that we interpret as consciousness. By describing "qualia" as perceived experiences by conscious beings, we are creating imaginary problems. This is an outcome of incorrectly believing that when an organism experiences things, there must be an entity existing apart from the brain and its mind, which becomes involved in perceiving the experiences that have already transpired. This involves an illogical separation of the experiential process into the experience and the experiencer. This is exactly the same division that occurs when thinking in terms of oneself and the one observing the self.

What we learn from our experiences enables us to adapt to changes in our world and then to modify our responses. This process is greatly facilitated when the brain realizes that it is the source of thoughts, perceptions, sensations, emotions feelings, and various experiences, etc., which ends the illusion of "consciousness" existing as a "thing."

Without the presence of a neural network, which is what the mind actually is, there would be no sense of experiencing. The subjectivity of the experience is an outcome of the specific neural patterns that exist in each brain, through which all information is processed and interpreted uniquely.

Sometimes a different question expresses the imaginary hard problem: How can any mental processes be conscious? This is easy to answer. Mental processes by themselves are not conscious. (Further discussion of this is in the section on HOR theories in this chapter.) Thoughts are not alive. Thoughts are products of brain activity. Because consciousness is not a thing, but a category for many mental processes and experiences, it is not something that the brain has. Referring back to the "running" analogy, it is incorrect to state that I have running. Running describes what occurs when a particular coordinated brain-and-leg-muscle activity is happening. Similarly, thinking describes what occurs when there is activity in the associative areas of the cerebral cortex.

Evolutionary Benefit of Advanced Mental Processes

The brain's output constituting the conscious and "unconscious" mind are quite adept at functions involving the automatic nervous system, such as breathing, in addition to performing physical skills that we learn, including precise athletic ones. However, the ability to gain useful new knowledge and skills by considering novel information and situations, and to solve difficult problems through innovative processing, requires higher-level programs and processes such as analyzing and interpreting information associated with the consciousness abstraction.[93]

Uniting different ideas based on the form of the information, discovery of underlying natural laws, and noticing the most significant deep structures within what enters our awareness, are very significant aspects of innovative cognition, which provides a profound learning tool and ability to create meaning.[94] The ability to recognize patterns facilitates great levels of understanding, which enables more control of our environment.

Specificity of brain processes results from specialist cortical processing regions, while networking of general-purpose areas combines many sources of very structured meaningful information from those regions, using extremely rapid brain rhythms, to generate mental phenomena associated with the consciousness concept.[95] Neuroscientists have proposed creating mathematical models using knowledge of neural patterns, information signature and method of communication of the identified areas of the brain to explain what occurs.[96]

Summary of Discoveries of Neuroscience

Neuroscientists found that all mental processes are equivalent to electrical activity of an enormous group of information-processing brain cells, each of which receive and transmit chemical signals termed neurotransmitters, with 1 indicating a firing, and 0 no firing, to thousands of other neurons, thereby influencing and being affected by their activity.[97] However, it takes a specific number of outputs from other neurons, which increases with mental fatigue, to elicit a firing response. Enhanced sensitivity among neuron groups facilitates greater efficiency.

HOR Theories

Higher-Order Representation (HOR), a.k.a. Higher-Order Thought (HOT) theories, provide a propitious account of consciousness in terms of mental states. These theories, which originated with David Rosenthal in 1986, arise from conceptual, methodological and phenomenological sources.[98] Conceptual motivation emerges from differences and similarities between ideas concerning what the term "consciousness" means. Phenomenological motivation focuses on studying the development of human consciousness and self-awareness within philosophy and results from HOR theory having a high regard for the subjectivity of conscious experiences. Because HOR theory has content in which a subject is included, it has a useful framework from which to explain this subjectivity.[99] Methodological motivation stems from HOR theories having resources to explain consciousness in a naturalistic way. For example, HOR theorists seek to elucidate a difficult to comprehend mental phenomenon by examining its aspects, such as intention. Because intention is conducive to explanations involving physical terms, this implies that consciousness is also.[100]

There are several strengths of HOR theories' explanatory approach. An important one is that intentionality links to phenomenology as indicated by studies in the philosophy of mind.[101] Another benefit is HOR theory could retain significant explanatory value without requiring physicalism to be valid. For example, if it is not possible to show a naturalistic explanation for intentionality, HOR theory could offer the tools to develop an explanation of consciousness that is not naturalistic.[102]

A third useful feature is their focus on providing an account of mental states that are conscious and on what determines this.[103] Although the concept of consciousness is ambiguous, there is a distinction between attributions of consciousness to mental states and to subjects. HOR theorists distinguish between state consciousness and creature consciousness. The latter, exhibited by all sentient and awake creatures, is further divided into transitive, which refers to being conscious or not, and intransitive, which means being conscious of something. Introspecting on one's conscious states is a transitive type of consciousness, because its objects are mental states.[104] Based on this understanding, we can state HOR theory as: "intransitive state consciousness is explained in terms of transitive consciousness of mental states."[105]

State consciousness, a feature of mental states, demarcates the difference between conscious and unconscious states. Conscious ones include visual perceptions, emotions and pain, etc., while unconscious ones include implicit beliefs and unacknowledged desires. HOR theorists attempt to explain how to distinguish mental states that are conscious from unconscious ones, by accounting for intransitive state consciousness with transitive creature consciousness terms. Conscious states refer to those that subjects are aware of being in.

HOR theorists contend that this approach facilitates ascertaining how one is conscious of their first-order state. Being predominantly concerned with modeling subjective aspects of consciousness, they provide a separate account of mental features, allowing mental qualities to exist in conscious or unconscious mental states.[106]

HOR proponents account for consciousness by a higher-order mental state taking another state, such as a sensation, as its object. Though this means the lower-order state is conscious.[107] Thus, according to this theory, the mystery of consciousness is an outcome of the nature of conscious states: mental states are conscious through higher-order representation of a mental state.[108]

HOR Solution to Hard Problem

In seeking to account for what causes a mental state to be a conscious mental state, HOR theorists propose the existence of an intentional relationship between the higher-order and lower-order states, wherein the higher-order state is "about" the lower-order state.[109] An example of an intentional state is perceptions, which can contain a great variety of sensory information. Another example is thoughts, which can be about real things, abstractions or other thoughts.

Characteristics of Higher-order Thought Theory

HOT theories attempt to explain the nature of conscious states as being the result of higher-order thought about them. It is important to understand five significant features of HOT theory. David M. Rosenthal wrote about one of these in pointing out that state consciousness has to have an immediacy about it as occurs when one is in pain.[110]

According to Rosenthal, another significant attribute of higher-order thoughts is that to distinguish between conscious and non-conscious states they transpire simultaneous with the represented mental state. In addition, the higher-order thought must assert that I am in a specific mental state, rather than hope or speculate about this.[111]

A third characteristic of HOT theories is that consciousness is a relational property of mental states; not inherent to them. By forming a representational relation to a higher-order thought, mental states are able to be conscious.[112]

A crucial need for self-reference, which requires a capacity to differentiate self from not self, within HOT content is a fourth component.[113]

A fifth element is that higher-order thoughts are not introspective states. Rosenthal stated that although higher-order thought enables the mental state to be conscious, the HOT itself is never conscious of anything, unless there is a higher-order thought about it, which produces a state of rare deliberate introspection.[114]

Wide-Intrinsicality View (WIV)

In this 1996 version of HOT developed by Rocco Gennaro, conscious states are comprised of a mental state and a higher-order thought concerning it, with consciousness being intrinsic to these complex states.[115] This feature distinguishes WIV from other HOT versions. Gennaro contends that for a mental state to become conscious does not require new relations external to it, but rather depends on changes to it. He posits that consciousness resides in the state constituted by a mental state and the higher-order thought concerning it; the higher-order relation is not external, but rather internal to the conscious state.[116]

A mental state's quality depends on higher-order relations. To create a consistent arrangement of objects when organizing sensory data, we have to apply concepts to sensory states.[117] Gennaro argues higher-order thought offers conceptual resources to determine the content of the conscious sensation; it is necessary to apply a higher-order concept with an implicit self-referential part to any sensation for it to result in a particular content for oneself, such as "hurtful."[118] In this way, sensations enter consciousness via higher-order thought.[119]

WIF contends that the intrinsic nature of consciousness is fully compatible with its complexity. Although the property of consciousness is intrinsic to an entire complex state, a conscious state is explicable in terms of the mental state and a higher-order thought comprising it.[120]

HOT Theory

In 2000, Peter Carruthers proposed a version of HOT in which higher-order thoughts function "dispositionally," which involves mental states being conscious once they are accessible to a brain.[121] He posited that two perceptual systems exist. One uses unconscious perceptual states, while being primarily action guiding. The other system produces beliefs about perceptual data, so perceptual states need to include information about the sensation and about one's experience of it.[122]

Carruthers maintains that the information carried by a mental state somewhat determines its content. Perceptual states are conscious when they acquire the higher-order content that constitutes the experiential aspect of the perceptual state.[123]

Because the belief-forming perceptual system's role is to produce higher-order thoughts, the states in this system acquire perceptual content, such as "a thing is painful," and the higher-order content, such as "a thing seems painful."[124] The subjective quality of "seeming painful" arises from one's thought concerning the pain and forming beliefs and attitudes about it. [125]

This HOT version is a form of higher-order perception theory. We perceive a green plant even though we do not know it is green and unable to specify its shade of green. Additionally, perceptual states take on higher-order content through being accessible to a brain able to generate higher-order thoughts, and they possess this dual content without requiring the formation of such a thought.[126] Therefore, the occurrence of consciousness does not depend on any actual higher-order thought or higher-order perception.[127]

Higher-order Perception Theory

This provides a higher-order account of consciousness with its origins with John Locke who linked consciousness with the reflective process: "Consciousness is the perception of what passes in a man's own mind." (*Essay II, 1, §19, 115*).[128] In *Consciousness and Experience* (1996), William Lycan contended that intentionality, functional organization, and specifically, second-order representation of one's own mental states, accounts for all of what we call consciousness or mind.[129] He refers to his idea as the "hegemony of representation."[130] This conforms to my view that consciousness is simply an abstraction.

This theory holds that when internal scanners produce perceptual representations of mental states, those states are conscious. Internal scanning is higher-order monitoring that enables one to plan and monitor one's action though the coordination and relay of information on mental states by producing higher-order perceptual representations.[131]

Evaluation

Fundamental to HOT theory is the idea that higher-order thoughts distinguish conscious from unconscious states, and "state consciousness" forms the mystery of consciousness. Although higher-order perception theory accounts for contrasts between these two states, its most significant strength resides in explaining, "what it's like for the subject" to be in conscious states rather than unconscious states.[132] Because HOT theories account for consciousness by an instinctual higher-order process without attentive self-consciousness, these theories do not have to incorporate undiscovered morphology or mysterious processes.[133]

HOT theorists explain that consciousness may be mysterious in some ways, but it is largely explainable: increasingly complex states link to less complex states and relations, such as "aboutness." Because it is included with all mental phenomena, it is therefore possible to account for it in physical, naturalistic terms.[134]

William G. Lycan contends that intentionality, functional organization, and specifically second-order representation of one's own mental states fully explains what is termed consciousness and mind.[135]

Some critics argue that because it appears higher-order states can carry out their actions without any conscious feeling, therefore the theory does not solve the "hard problem of consciousness." However, proponents respond that they can explain qualitative differences, such as exists between colors, by variations in first-order sensory states. They assert that higher-order representation is just what consciousness is, and a conscious state occurs when one is conscious of a mental state.[136]

Linkage between finer sensory discernment and concept attainment bolster a higher-order-thought accounting of consciousness. Lycan proposes that everyone has a particular way of representing their mental states, which derives from the functions that higher-order representations have within the representational system known as the mind.[137] Sharing referents, such as something's color, enables us to converse with others and account for aspects of our first-order sensory representations. We exit the language domain, leaving only individual ways to refer to our sensations, in proceeding to higher-order representation.

HOT theorists do not intend it to analyze the logical relations for the concept "consciousness," but rather to use it as an observational supposition. Its proponents see its purpose as creating a more substantial structure to comprehend the mind by accounting for how consciousness relates to thought, perception, and other mental states.[138] My criticism of HOT and HOR theories stems from my defining consciousness as an abstraction for cognitive and emotional processes. Using this definition, I disagree with the statement that "a conscious state occurs when one is conscious of a mental state." In my thinking, the mental state is always within the category of consciousness. All states of the brain are conscious states, although some get more attention from the brain. I would argue that while dreaming, the imagery occurring in the brain is as much a part of consciousness as the waking state. Additionally, consciousness is still present in the deep sleep referred to as slow-wave sleep, but it is of a different quality.

It is important to understand that there is no separate existing or occurring "phenomenon" of consciousness. Simply stated, consciousness is the motion of neurons in structured patterns. There is no entity that is conscious of this process. There is only the process. Even the thought that "I am conscious," is still the result of this process. That is a very liberating thing for the brain to understand, because it eliminates the cause of all inner conflict, which is the sense of duality between the "thinker" and the "thought."

Computational and Representational Theories of Mind

The computational idea, which originated with Hilary Putnam in 1961, is presumed by evolutionary psychologists, and is widespread in cognitive psychology. During the 1960s through the 1980s, Jerry Fodor,[139] a cognitive scientist and philosopher developed it further. The theory sees the brain as an information processing system, with thought as a type of computing. The mind is the output of the computation and results from the program the brain runs. To compute real objects, the "inputs" represent objects. Thus this theory needs mental representation, because the brain can only compute representations of things.

In Fodor's view mental states are computational relations that we hold to mental representations that the brain realizes.[140] He contends that analogous to the sentences in a language, in which words and sentences are put together based on grammatical rules, these representations also contain compositional semantics and are syntactically structured.[141]

In my estimation, the most powerful program running is the one I label as the "self," which provides input representations to generate the output representations of reality. The representational theory contends that mental representations function as intermediaries between objects or processes being observed and the observer. These intermediaries represent objects to the mind. Differing from the computational theory, this approach maintains that every mental state, including those involving qualia is representational and thus can be computed in the brain.

Self as Collection of Programs

Because it is not possible for a program to be conscious, the "self" cannot have knowledge of itself. The brain can have perception of self, but self cannot have knowledge of itself. The brain receives inputs from its senses, which are interacting with the environment. This process results in outputs in the form of thought, which includes all imagery and auditory representations, emotions, feelings and actions. The brain knows it is aware of these outputs.

This output becomes the source of secondary inputs, not from the environment, but rather from the brain itself, which the brain can be aware of. However, once again it is important to clarify that this does not mean the brain is "aware of its own awareness." It is simply being aware of its own outputs that are being treated as inputs in considering them. In an interview on *Consciousness Central 2017* Noam Chomsky demonstrates the identical understanding as my own in his statement: "The internal workings of the mind are inaccessible to consciousness and even when we produce what is called inner speech... that's the external system. It's not getting at what's going on internally."[142]

As a result of processing its inputs, and the outputs which form inputs upon the brain's awareness of them, the brain can attain greater understanding based on how accurate and thorough it is in this function. How many rounds this process takes is limited by the individual's mental abilities, experience in this process, and to what depth it appears necessary in living intelligently.

The more accomplished one is in this process, the quicker it occurs, without any noticeable effort. One of the outcomes of exploring in this field of awareness is that the brain becomes far more sensitive and clear. It stops seeing itself as cut off from everything else, including life forms. As this different perspective develops a new ability arises in the brain. This is genuine true compassion. Although many people can empathize very well with others, compassion is completely different, because it does not involve trying to imagine oneself in someone else's situation. On a very deep level, there is simply a sense of not being separate from others, even though one maintains awareness of one's physical existence.

References

---

[1] Mitchell, J. *Insight into the Seat of Human Consciousness.*

[2] Fox, M.D, cited by Mitchell, J. *in Insight into the Seat of Human Consciousness.*

[3] Ibid.

[4] Mitchell, J. *Insight into the Seat of Human Consciousness.*
[5] Ibid.
[6] Mitchell, J. *Insight into the Seat of Human Consciousness.*
[7] Nappi, Bruce, from e-mail to me.
[8] Bor, *The Ravenous Brain*, 163.
[9] Parsons, Science 1001, *Absolutely Everything That Matters in Science in 1001 Bite-Sized Explanations.* 295.
[10] Bor, *The Ravenous Brain: How the New Science of Consciousness Explains Our Insatiable Search for Meaning*, 7.
[11] Wikipedia, Category Mistake.
[12] Borchardt, *The Scientific Worldview*, 249.
[13] Evans, J, Two Minds: Dual Processes and Beyond.
[14] Iid.
[15] Sage, J. *Mind, Brain and Consciousness. Do you believe the mind is identical to the brain.*
[16] Lui, J. H.; Hansen, D. V.; Kriegstein, A. R. <u>Development and Evolution of the Human Neocortex.</u>
[17] Lodato, S.; Arlotta, P. *Generating Neuronal Diversity in the Mammalian Cerebral Cortex.*
[18] Baars and McGovern, *Global Workspace Theory.*
[19] Wikipedia, Global Workspace Theory.
[20] Baars *The global brainweb: An update on global workspace theory.*
[21] Wikipedia, Global Workspace Theory.
[22] Baars *The global brainweb: An update on global workspace theory*
[23] Ibid.
[24] Baars and McGovern, *Global Workspace Theory.*
[25] Baars *The global brainweb: An update on global workspace theory*
[26] Baars, *A Cognitive Theory of Consciousness.*
[27] Wikipedia, *Global Workspace Theory.*
[28] Ibid.
[29] Ibid.
[30] Baars *The global brainweb: An update on global workspace theory.*
[31] Wikipedia, chunking.

[32] Ibid.
[33] Bor, *The Ravenous Brain*, 138.
[34] Ibid.
[35] Ibid.
[36] Llinás, *Consciousness and the Brain*.
[37] Wikipedia, **t**halamus.
[38] Bor, *The Ravenous Brain*, 112.
[39] Ibid., 135.
[40] Ibid., 113.
[41] Ibid., 115.
[42] Ibid., 115.
[43] Ibid.,116.
[44] Ratey, *A Users Guide to the Brain*, 111.
[45] Ratey, 112.
[46] Ibid., 111.
[47] Ibid.,129.
[48] Ibid., 132, 133.
[49] Ibid., 133.
[50] Father Vincent. Quoted in Philosophy of Freewill/ Human Functional Freewill.
[51] Baron d'Holbach, *Quotes*.
[52] Moore, *Determinism and Free Will*.
[53] Smith, *Neuroscience vs philosophy: Taking aim at free*.
[54] Soon, Brass, Heinze[4] & Haynes[1,2] *Unconscious determinants of free decisions in the human brain*.
[55] Bode, He, Soon, Trampel, Turner, and Haynes, *Tracking the Unconscious Generation of Free Decisions Using Ultra-High Field fMRI*.
[56] Fried, Mukamel and Kreiman, *Internally generated preactivation of single neurons in human medial frontal cortex predicts volition*.
[57] Smith quoting Fried, *Neuroscience vs philosophy: Taking aim at free*

will.
[58] Smith quoting Mele, Ibid.
[59] Ibid.
[60] Ibid.
[61] Smith quoting Roskies, Ibid.
[62] Smith quoting Glannon, Ibid..
[63] Ibid.
[64] Smith quoting Gazzaniga, Ibid.
[65] Wiese, W. and Metzinger, T. Philosophy of Mind and Predictive Processing.
[66] Ibid.
[67] Ibid.
[68] Ibid.
[69] Merriam-Webster, *Definition of* Self-Awareness.
[70] Parsons, *Science 1001*, 295.
[71] Wikipedia, Damasio's Theory of Consciousness.
[72] Parvizi and Damasio. Consciousness and the Brainstem.
[73] Wikipedia Damasio's_theory_of_consciousness.
[74] Ibid.
[75] Ibid.
[76] Merriam-Webster, *Definition of* Nirvana.
[77] Chalmers, *Facing Up to the Problem of Consciousness*.
[78] Ibid.
[79] Ibid.
[80] Ibid.
[81] Chalmers, *Facing Up to the Problem of Consciousness*.
[82] Ibid.
[83] Ibid.
[84] Ibid.
[85] Weisberg, *Hard Problem of Consciousness*.
[86] Ibid.
[87] Ibid.

88 Ibid.
89 Ibid.
90 Ibid.
91 Ibid.
92 Borchardt quoting in *The Scientific Worldview*, 251.
93 Bor, *The Ravenous Brain*, xiv.
94 Ib*id., xv.*
95 Ibid., xvi.
96 Ibid., xvi.
97 Bor, The Ravenous Brain, 11.
98 Bruno, *Review of: Gennaro, Rocco J.(ed.) Higher-Order Theories of Consciousness: An Anthology.*
99 Ibid.
100 Ibid.
101 Ibid.
102 Ibid
103 Ibid.
104 Droege, *Higher-Order Theories of Consciousness.*
105 Ibid.
106 Ibid.
107 Ibid.
108 Ibid.
109 Ibid.
110 Ibid.
111 Ibid.
112 Ibid.
113 Ibid.
114 Ibid.
115 Ibid.
116 Ibid.

[117] Ibid.
[118] Ibid.
[119] Ibid.
[120] Ibid.
[121] Ibid.
[122] Ibid.
[123] Ibid.
[124] Ibid.
[125] Ibid.
[126] Ibid.
[127] Ibid.
[128] Ibid.
[129] Ibid.
[130] Lycan, *Consciousness and Experience.*
[131] Droege, *Higher-Order Theories of Consciousness.*
[132] Ibid.
[133] Ibid.
[134] Lycan, *Consciousness and Experience.*
[135] Droege.
[136] Ibid.
[137] Droege, citing Lycan.
[138] Ibid.
[139] Fodor, J. Internet Encyclopedia of Philosophy.
[140] Ibid.
[141] Ibid.
[142] Chomsky, N. *Consciousness Central 2017* interview by Nick Day.

Chapter 15

Summary and Implications

Sir Isaac Newton wrote: "To explain all nature is too difficult a task for any one man or even for any one age. Tis much better to do a little with certainty and leave the rest for others that come after than to explain all things by conjecture without making sure of any thing."[1] Those words show much wisdom. Considering that they originated with arguably the greatest scientist, the statements also indicate great humility, an indispensable quality for anyone looking to discover the nature of reality.

Notfinity Process challenges the dominant worldview in physics, cosmology and consciousness studies, all of which have many unfounded assumptions and invalid abstractions. In addition to my ideas on consciousness being an abstraction for brain processes, rather than a thing or process in itself, and my views on theoretical physics, this book presented several alternative theories to the standard models. These include David Bohm's realist interpretation of quantum mechanics experiments, Steven Bryant's modern mechanics, Glenn Borchardt's infinite Universe theory and Duncan Shaw's aether gravity model. Unlike QMT and the two relativity theories, these do not have contradictions or require that we accept impossible concepts.

The ideas presented herein, such as univironmental determinism, have profound implications and significance for how we see reality and ourselves. For example, the invalidating of the Big Bang Theory means finity is impossible, because it requires a beginning.

The paradigm of an indivisible unified causal relationship to the Cosmos supersedes the dualistic one. *Notfinity Process* differentiated consciousness from awareness and perception and considered the problems caused by Descartes' dualism with its associated illusory idea of "free will." It examined Antonio Domasio's model, HOR/HOT theories, Baar's and McGovern's GWT, and the "hard problem," as they relate to discoveries of neuroscience.

The book explored the pervasive illusion that a self exists apart from the products of the brain's programs and processes. It challenged what most people assume as an axiom, namely that the "self" is in control of the brain and behavior. In proposing that self is not the operative agent in behavior, I presented studies that showed the brain decides things before there is consciousness of the decision. Additionally I discussed the theory that the brain is a biological computer, with the software being the self and the output being the "mind." This presents a more scientific understanding than the commonly held idea that a self exists apart

I discussed concepts of self-awareness and revealed the reason that it is impossible for the self to be aware of itself. If this understanding becomes widespread amongst therapists and clinical counselors, it can redirect their attention to helping their clients improve their situation rather than trying to engender a mythical higher self-awareness. I finished up by briefly explaining Jerry Fodor's computational theory of the brain and mind, including my understanding of the self as a visualized experience using memory and programs, and the role of hormones in the brain process termed emotions.

Nofinity Process showed that we are not in any way separate from the natural world of universal causality. Therefore, our behavior and development completely depends on environmental and genetic factors, with no role for an uncaused will. This perspective facilitates the discovery of real causes of behavior, which leads to a deeper understanding of what type of environments contribute to positively functioning individuals and thus better societies.

Realizing that our behavior and psychological characteristics have external causes brings a change in attitude and methods of dealing with destructive antisocial behavior. Rather than focusing blame on individuals as the cause of their behavior through the supposed exercising of a hypothesized "will," we can direct greater attention to eliminating the conditions that contribute to producing the undesirable actions. This can result in more positive social policies founded on the knowledge that all behavior is causal.

Summary and Implications

\Although we are all unique physically and psychologically, our individual existence is part of a cosmic expression. Seeing ourselves as inextricably related to nature and part of it is a completely materialistic perspective that engenders a deep sense of oneness. Rather than having our significance depend on an imagined impossible immaterial world, where one's life supposedly continues after it ends, meaning in one's life arises from the quality of the relationships we create, and whatever contributions we make to society and our macrocosm.

Having seen the problems with the orthodox theories discussed in this book, you may have concluded that these models in physics, cosmology and consciousness are invalid. The alternatives presented in this book provide a perspective that has the potential to lead to greater understanding.

In evaluating any novel theory such as Bryant's Modern Mechanics, what criterion is best to use? The best response is to focus on the mathematical results that they produce. Those that give superior predictions to the orthodox models deserve more examination.

If this book has increased your appreciation of reality, and brought on a deeper comprehension of the nature of our inseparable connection to the notfinity process of microcosms in motion, then it has served a useful purpose. Let us work together to create a positive world society.

References

---

[1] Newton, cited in *New Mexico Museum of Space History*.

Definitions

Aether. All forms of ordinary matter (a.k.a baryonic) smaller than electrons, which divide infinitely. Thus, larger aether forms consist of smaller sized aether. It is not an elementary particle; no such objects exist in Neomechanics.[1]

Awareness. This is a noun for the outcome of the process of the brain taking in raw, unfiltered sensory input, and what occurs as a result of the thinking and emoting process.

Baryonic matter. In astronomy, it refers to all objects, which are ordinary atomic matter. essentially ignoring the presence of electrons which, after all, represent only ~0.0005 of the mass.[2]

Complexification A gravitation process that changes the parts of any kind of matter into complexes of closely bonded matter.[3]

Consupponible. Applied to two or more non-contradictory assumptions where one of them implies the other's validity.[4]

Convergence. "The act of converging and especially moving toward union or uniformity."[5]

Copenhagen interpretation. A QMT view positing that measurements influence systems, resulting in a wave function collapse. Links with the concept that before doing a measurement, physical systems lack definite properties and it is only possible to make predictions on probabilities that specific measurements produce specific outcomes. Key elements are indeterminism, immaterialism, lack of full causality, and uncertainty principle.

Cosmogony. 1. This is any theory involving the origin of the universe. 2. The creation or origin of the world or universe."[6]

Cosmology. "1 a: a branch of metaphysics that deals with the nature of the universe b: a theory or doctrine describing the natural order of the universe. 2: a branch of astronomy that deals with the origin, structure, and space-time relationships of the universe; also: a theory dealing with these matters.[7]

Cosmological Constant. Λ. A term used in relativity gravity equations for a repulsive force, which may partly cause the universe's expansion rate.[8]

Determinism. Perspective that causality is universal; all events have physical causes including behavior, thoughts and choices.

Dark Energy. In orthodox physics, it is a hypothetical energy infusing all space, and is the energy density of the "vacuum." Some physicists claim cosmological constant and scalar fields are forms of dark energy, which is equivalent to vacuum energy.[9] In IUT this represents large-scale motions of matter.[10]

Dark Matter. Cosmologists speculate that it is non-baryonic, and infer its existence to provide sufficient total matter and gravitation to keep galaxies together. In UCT Puetz and Borchardt theorize that it is non-luminous baryonic matter.[11]

Domain. Input into the function. Often refers to the set of all real numbers since many mathematical functions can accept any input.

Doppler Shift. The shift in the received frequency and wavelength of an electromagnetic or sound wave when the observer or source is moving.

Dualism. "A theory that considers reality to consist of two irreducible elements or modes."[12]

Eigenvalue. "A scalar associated with a given linear transformation of a vector space and having the property that there is some nonzero vector which when multiplied by the scalar is equal to the vector obtained by letting the transformation operate on the vector."[13]

Energy. A matter-motion term concerning the exchange of matter's motion[14] representing a calculated result from a number for mass times the square of a velocity number.[15]

Equivalence principle. A generalization of the principle of relativity

Definitions

basic to GRT. It states that there is no difference between gravitation and accelerated motion.

Fatalism. This is an idealistic perspective that denies that a human microcosm has the matter in motion within it to affect its macrocosm."[16] This idea excludes the univrironmental concept.

Field. An effect caused by aether matter circulating around the center of a vortex.[17]

"Free Will". The supposed capacity to decide between alternative courses of action in which the outcome has not been causally determined. The doctrine that this ability can be independent to some degree of material causation[18] often based on the belief in immaterial consciousness separate from a body that does not result from matter in motion.

Function. An equation for which any x entered into it results in one y. Here is a relation that is a function: {(1,5) (5-2) (3-4) (6,0) (9-3) (4,7) (8,5)} Making two groups, one each of the first two numerals produces: {1,5,3,6,9,4,8} and { 5,-2, -4, 0,-3,7,5}. This is a function because only one ordered pair begins with any number from the first set. Modern mechanics uses functions as developed in computer science. In MMT, a function definition is an expression in the form: function_name(parameter_list)={function_body}"[19] Bryant gives an example function definition: $f(x) = \{x2\}$. f is the functions name, x2 is the body or equation inside these brackets { }, and the name plus any of the function's variables or symbols as parameters is the signature.[20] After defining a function, before using it, one invokes it, which means inserting numbers in place of the parameters. A function invocation (a.k.a. instantiation) takes this form: instance_variable=function_name (argument_list).The instance variable is where the instantiated expression goes. Arguments or an argument list refers to variables or numerals in the global namespace that replaces a function's parameters when it is invoked, and arguments can be expressions made from variables and numerals on the namespace.[21]

General Relativity Theory (1915). This describes gravitation at very large levels in which mass or energy supposedly accounts for it by curving hypothesized four-dimensional spacetime. This theory includes a set of equations that determines the amount of curvature produced by any specified mass or energy.

Geometric transformation. A function whose domain and range are sets of points, which are usually both R^2 or both R^3, and are often required to be 1-1 functions, so that they have inverses.[22]

Global Workspace. This contains imagery, inner speech and subjectively experienced events held only a couple of seconds, rather than the 10 to 30 of working memory.[23]

Heisenberg uncertainty principle. This derives from a concept that all matter and energy behaves similar to a particle and a wave. This results in a fundamental precision limit, for which specific pairs of a particle's physical properties, such as its momentum and position, are knowable. Hence, increases in accurately determining one property reduce our certainty for the other property. This also applies to fields. One cannot make precise measurements of their value and their rate of change. The identical problem exists for many other pairs of quantities.

Higher Order Representation (HOR) or Higher Order Thought (HOT) theories:. "Consciousness" is accounted for by higher-order mental states taking lower-order states, such as a sensation, as its object, which enables the latter to be conscious..

Holomovement. This unites Bohmian physics ideas that all matter is in a universal flux or becoming process, with the holistic principle of undivided wholeness, which is a dynamic wholeness-in-motion with everything moving together in an interconnected process.

Homogenous. This refers to having identical properties at every point; and being uniform without irregularities.[24]

Hubble ultra Deep Field (HUDF). This is an image of a small region of space in the Formax constellation with 10,000 galaxies

Idealism. "Any system that reduces all existence to mind or thought. This may be either a single, absolute mind or thinker or a plurality of minds."[25]

Immaterialism. "A philosophical theory that material things have no reality except as mental perceptions."[26]

Inertia. "A property of matter by which something that is not moving

Definitions

remains still and something that is moving goes at the same speed and in the same direction until another thing or force affects it."[27]

Instantiate. To represent an abstraction by a concrete physical thing.

Isotropic. Uniformity in every orientation. "Equal in all directions."[28] In cosmology, it refers to no preferred direction.

Macrocosm. A specific microcosm's external universe.[29]

Materialism. A theory that all processes and being arise from manifestations or results of matter, which is the only reality.[30] It assumes that the Universe continues to exist without observers.[31]

Matter-Space continuum. "A range or series of microcosms that are slightly different from each other and that exist between what we imagine to be perfectly solid matter and perfectly empty space."[32]

Mechanism. A theory that matter in motion accounts for all phenomena and can be explained using mechanics.[33]

Mega-vortex This is a hypothetical massive spiraling structure bigger than the observable universe.[34]

Metallicity. The fraction of mass of a star or other kind of astronomical object that is not hydrogen or helium.[35]

Microcosm. All things and portions of the Universe. It replaces the concept of bodies, objects and systems.[36]

Motion. An occurrence that describes what microcosms do.[37]

Nadal. A noun for the measurement of amounts of motion as indicated by clocks, rather than the actual movement, designated by "time." It represents the Earth's rotational motion. Rather than referring to the moving that is occurring, it specifies the movement as indicated by an object's positional change between each measurement. By measuring the motion of one microcosm's change in position relative to another microcosm, it is possible to ascertain a specific time, which since the first time-keeping devices has used the Earth's rotation as the comparison standard.

Nadalting. A verb for measuring the motion that is "time," which refers to moving.

Neomechanics. An improved and modified classical mechanics, assuming infinite causes.[38]

Notfinity. This involves the absence of any possibility of finity, rather than an abstraction of infinity.

Orthodox physics. This refers to relativity and quantum mechanics theories.

Platonia. Nows existing simultaneously as points in a timeless, unchanging universe. A word coined by Julian Barbour.

Principle of Relativity. This is the foundational principle of SRT. It states that an identical set of physical laws hold for all observers having a constant velocity, and thus all constant velocity observers can justifiably maintain that they are at rest. Motion for any object occurs only in comparison or relation to another object. The principle of equivalence generalizes this principle.

Process. What occurs from microcosms continually moving.

Qualia. Subjective perceived experiences.

Quantum mechanics theory. This describes things of atomic size and smaller. All waves are emitted and observed in packages known as quanta, which cannot be subdivided. For example, light is quantized as photons. The uncertainty principle is part of this theory. Referring to this aspect, Stephen Hawking stated that QMT is "a theory of what we do not know and cannot predict."[39]

Range. This is the output from the function.

Redshift. Spectral shift in visible light's color from a galaxy when any EMR, including light, increases its wavelength to the red end of the spectrum equivalent to reduced photon energy and frequency. Orthodox cosmologists interpret it as being caused by the Doppler effect, gravitational effects and cosmic expansion, which is erroneously presumed.[40]

Refraction. This refers to a propagated wave's change in direction resulting from a change in medium of transmission.

Relation. A set of ordered pairs.

Self-awareness. Having knowledge of one's personality, character, or

Definitions

individuality, and being able to introspect and discern oneself from others and the environment. This includes the concept of the brain perceiving the visualized experience known as "self."

Spacetime. A matter-motion noun used by indeterminists seeking to unify and objectify space and time, which is conceptually impossible..[41]

Space-time. A deterministic matter-motion adjective that relates a thing's location to the motions of all things,[42] as described in the Infinite Universe Theory.

Special Relativity Theory (1905). A view of space and time which holds that scientific laws are identical for all freely moving observers irrespective of their speed. Regardless of the velocity of the observer who measures the speed of light, it remains constant.

Submicrocosm. Part of the Universe within another part.[43]

Supermicrocosm. Portion of the Universe external to another part.[44]

Superposition. A QMT concept that things supposedly exist in every possible state simultaneously until someone makes an observation or a device takes a measurement of a thing that then causes it to be in just a single or a smaller set of states.

Systems Theory (Philosophy). Belief that any part of the Universe may be isolated from the rest and studied as a system. It gives more attention to internal interactions over external ones. It is the present dominant scientific worldview.[45]

Thing. "The concrete entity as distinguished from its appearances."[46] An object.

Timing. A verb, equivalent to the word moving. The Earth is moving or rotating is the same as the Earth is timing. "Universal time is the motion of each portion of the universe with respect to all other portions. In practice we obtain local time by measuring the change in location of one thing with respect to another"[47]

Translation. This is a mathematics term for preserving orientation in addition to congruence.

Universe. This concept attempts to represent the existence of unlimited microcosms of an infinite variety of sizes, and to the occurrence of their limitless motions and interactions with their macrocosm and with other microcosms. However, because no ultimate largest structure or complete collection of microcosms exist, nor is there a volume boundary for them, "Universe" cannot refer to a thing or object because these are limited. This word represents an abstraction that cannot be instantiated. As there is no ultimate largest structure or complete collection of matter and no volume boundaries, this word cannot refer to a thing or microcosm because all things and microcosms are limited. This word is only an abstraction.

Univironment "A microcosm plus its macrocosm. Every univronment is absolutely unique at all times."[48] "…that combination of the matter in motion within the microcosm and the matter in motion in the macrocosm that is responsible for the motion of the microcosm."[49]

Univironmental determinism. Every microcosm's evolution depends completely on the motions of matter within and without, and proceeds toward univironmental equilibrium. Because UD derives from causality, it assumes that there are an infinite number of causes for all effects.[50]

Wave. A pattern of matter spread out over a volume of space rather than localized areas. Examples are water, sound and light waves. In QMT it only designates presence of propagations, frequencies, interference, and their properties usually associated with a wave.

Wavelength. Distance between two successive corresponding points, such as two crests of a wave in a single cycle.

WMAP. The Wilkinson Microwave Anisotropy Probe. A satellite that measures the temperature found in space.

Worldview. This is the way one views life or contemplates the world and involves a particular world concept or philosophy of life of an individual or a group. It is "an articulated system of philosophy or a more or less unconscious attitude toward life and the world."[51] It includes a quality of being comprehensive, all-inclusive or fundamental in a wide range of areas.

References

---

[1] Puetz and Borchardt, *Universal Cycle Theory*.

Definitions

[2] Cosmos- THESAO Encyclopedia of Astronomy.
[3] Ibid.
[4] Borchardt, *The Scientific Worldview*, 341.
[5] Merriam-Webster, Definition of *Convergence*.
[6] Merriam-Webster, Definition of *Cosmogony*.
[7] Merriam-Webster, Definition of *Cosmology*.
[8] Merriam Webster, Definition of *Cosmological Constant*.
[9] Wikipedia, *Dark energy*.
[10] Puetz and Borchardt, *Universal Cycle Theory*, 12.
[11] Ibid, 12.
[12] Merriam-Webster, Definition of *Dualism*.
[13] Merriam-Webster, Definition of Eigenvalue.
[14] Borchardt, Do Spent batteries have more mass?
[15] Puetz and Borchardt, *Universal Cycle Theory*, 12.
[16] Borchardt, *The Scientific Worldview*, 305.
[17] Puetz and Borchardt, *Universal Cycle Theory*, 13.
[18] Wikipedia. *Free Will*.
[19] Bryant, Disruptive, 128.
[20] Ibid. 128.
[21] Ibid., 129.
[22] Wikipedia, Geometric Transformation.
[23] Wikipedia, Global Workspace.
[24] Wikipedia, *Homogenous* modified.
[25] Borchardt, *The Scientific Worldview*, 343.
[26] Merriam-Webster, *Definition of Immaterialism*.
[27] Merriam-Webster, *Definition of Inertia*.
[28] Borchardt, *The Scientific Worldview*, 343.
[29] Puetz and Borchardt, *Universal Cycle Theory*, 4.
[30] Borchardt, *The Scientific Worldview*, 343.
[31] Puetz and Borchardt, *Universal Cycle Theory*, 15.
[32] Borchardt, Infinite Divisibility of Matter and Space.

[33] Borchardt, The Scientific Worldview, 343.
[34] Puetz and Borchardt, *Universal Cycle Theory,* 15.
[35] Wikipedia, *Metallicity*.
[36] Puetz and Borchardt, *Universal Cycle Theory,* 15.
[37] Ibid., 15.
[38] Ibid., 16.
[39] Hawking, in Ferguson's *Quest for a Theory of Everything,* 29
[40] Ibid., 16.
[41] Borchardt, *The Scientific Worldview*, 344.
[42] Ibid. 344.
[43] Ibid. 344.
[44] Ibid.,344.
[45] Ibid.,344.
[46] Merriam-Webster. Thing.
[47] Puetz and Borchardt, *Universal Cycle Theory*, 18.
[48] Ibid.,18.
[49] Borchardt, The Scientific Worldview, 124.
[50] Borchardt, The Scientific Worldview, 157.
[51] Funk, What is a Worldview, cites Hunter Mead. *Types and Problems of Philosophy.*

Bibliography

Books

Barbour, Julian. *The End of Time: The Next Revolution in Physics.* Oxford University Press, Oxford, N. Y. (1999). ISBN 0-19-514592-5, pp.2, 44, 45, 49, 69, 96.

Bohm, David. *Causality and Chance in Modern Physics.* University of Pennsylvania Press, Philadelphia (1996) ISBN 0-8122-1002-6. Originally published in England by Routledge and Kegan Paul, London, 1957, pp 2, 170

Bohm, David and Hiley, Basil. *The Undivided Universe: An ontological interpretation of quantum theory.* Routledge, London, England (1993) ISBN 0-41506588-7, pp. 35-37, 382.

Bor, Daniel. *The Ravenous Brain: How the New Science of Consciousness Explains Our Insatiable Search for Meaning.* Basic Books A Member of the Perseus Books Group, New York (2012) ISBN 978-0-465-02047-8, pp. xiv, xv, xvi, 7, 112, 113, 115, 116, 119, 121, 135, 163.

Borchardt, Glenn. *The Scientific Worldview: Beyond Newton and Einstein.* iUniverse, Lincoln, NE (2007). ISBN 0-595-39245-9, pp. xxvi, 5, 10, 32, 36, 61, 62, 121, 122, 124,125, 188, 189, 190,to 197, 249, 251 305 320, 343, 344.

Borchardt, Glenn. *The Ten Assumptions of Science: Toward a New* Oxford University Press, Oxford, N. Y. (1999). ISBN 0-19-514592-5, pp.2, 44, 45, 49, 69, 96.

Bor, Daniel. *The Ravenous Brain: How the New Science of Consciousness Explains Our Insatiable Search for Meaning*. Basic Books A Member of the Perseus Books Group, New York (2012) ISBN 978-0-465-02047-8, pp. xiv, xv, xvi, 7, 112, 113, 115, 116, 119, 121, 135, 163.

Borchardt, Glenn. *The Scientific Worldview: Beyond Newton and Einstein*. iUniverse, Lincoln, NE (2007). ISBN 0-595-39245-9, pp. xxvi, 5, 10, 32, 36, 61, 62, 121, 122, 124,125, 188, 189, 190,to 197, 249, 251 305 320, 343, 344.

Borchardt, Glenn. *The Ten Assumptions of Science: Toward a New Scientific Worldview*. Lincoln, NE, iUniverse (2004). ISBN 0-595-31127-X, pp. 10, 15, 16, 21, 26 -28, 35, 47, 57, 60, 63, 76, 79, 87, 88, 91, 98, 10-108.

Bryant, Steven. *Disruptive: Rewriting the Rules of Physics*. Infinite Circle (2016). El Cerrito, CA. ISBN 978-0-9962409-0-1, pp. 13-29, 30, 31-35, 40, 52, 54, 67, 85, 93, 99, 102, 120, 123, 162, 203, 216, 223, 238, 251, 253, 256 -271, 281, 282, 288

Cole, K.C. *The Hole In The Universe: How Scientists Peered over the Edge of Emptiness and Found Everything*. A Harvest Book, Harcourt, Inc. San Francisco (2000) ISBN 0-15-601317-7, 170, 171.

Einstein, Albert. *Relativity The Special and the General Theory*, Introduced by Nigel Calder. Penquin Books Ld., London, England (2006), xxii. ISBN 978-0-14-303902-2., 45. 69, 70.

Evans, Jonathon. Two Minds: Dual Processes and Beyond. Oxford University Press, Oxford 2009. ISBN-13: 978-0199230167, ch 2 .

Feynman, Richard. Probability and Uncertainty—the Quantum Mechanical View of Nature, Chapter 6 of *The Character of Physical Law*, 129. Nov 9 to 19, 1964. Retrieved July 10, 2016.

Bibliography

http://www.informationphilosopher.com/solutions/scientistsfeynman/probability_and_uncertainty.html.

Feynman, Richard P. *The Character of Physical Law—6—Probability and Uncertainty*. Retrieved July 9, 2016Ferguson, Kitty. *Quest for a Theory of Everything*. Bantam Books, New York (1992). ISBN 0-553-29895-X., 29

Greene, Brian. *The Fabric Of The Cosmos: Space, Time, And The Texture OF Reality*. Vintage Books. A division of Random House, Inc., N.Y. (2004). ISBN 978-0-375-72720-,286. Pp.321-323.

Gribbin, John. *In the Beginning: The Birth of the Living Universe*. Little, Brown and Company (1993). Boston, New York, Toronto, London. ISBN 0-316-32836-7, 36, 37.

Guth, Alan. *The Inflationary Universe: The Quest For A New Theory of Cosmic Origins* (1997). Helen Books Addison-Wesley Publishing Company, Inc. Reading Massachusetts. ISBN 0-201-14942-7, xiii.

Heisenberg, Werner. *Physics and Philosophy, the Revolution in Modern Science*. Ruskin House, George Allen & Unwin Ltd., Museum Street, London. (1959) ISBN 04 J30016 x, 43, 67, 77.

Herbert, Nick. *Quantum Reality: Beyond The New Physics*. Anchor Books Doubleday, (1985)New York, London, Toronto, Sydney, Auckland. ISBN 0-385-23569-0, 27.

Kaku, Michio. *Hyperspace: A Scientific Odyssey Through Parallel Universes, Time Warps, and The Tenth Dimension*. Oxford University Press, Oxford, New York. (1994) ISBN 0-19-508514-0195, 196, 202.

Lightman, Alan. *Great Ideas In Physics: The Conservation of Energy, The Second Law OF Thermodynamics, The Theory of Relativity, And Quantum Mechanics*. McGraw Hill, New York. (2000) ISBN 0-07-135738-6, 44.

Marmet, Paul. *Absurdities in Modern Physics: A Solution Or: A Rational*

Interpretation of Modern Physics. Online edition http://www.newtonphysics.on.ca/heisenberg/chapter1.html.

Ratey, John J. A Users Guide to the Brain: Perception, Attention and The Four Theatres of The Brain. Pantheon Books, New York, 2001.
ISBN 0-67945309-1.

Parsons, Paul. Science 1001, *Absolutely Everything That Matters in Science in 1001 Bite-Sized Explanations.* Firefly Books, Hong Kong, China, 2014. ISBN 13:978-1-77085-501-4, 295.

Puetz, Stephen and Borchardt, Glenn, *Universal Cycle Theory: Neomechanics of the Hierarchically Infinite Universe.* Outskirts Press, Denver Colorado, 2011. ISBN 978-1-4327-8133-0 55995, 12, 13, 15, 16, 51, 82, 97-99, 213.

Ratey, John J. *A Users Guide to the Brain: Perception, Attention and The Four Theatres of The Brain.* Pantheon Books, New York, 2001. ISBN 0-67945309-1.

Reiter, Eric S. *Experiment and Theory removing all that Quantum Photon Wave-Partiicle Duality Entanglement Nonsense.* pp. 170 to 179. 2017 Proceedings John Chappell Natural Philosophy Society. July 19, 2017.Caledonia, Michigan.

Weinberg, Steven. *Dreams of a Final Theory.* Vintage Books, A Division of Random House, Inc., New York, (1992, 1993)), 67.

Zitzewitz, Paul, Neff, Robert F., Davids. Mark. *Physics Principles and Problems.* McMillan and McGraw, (1995) New York. ISBN 0-02826721-4, 155. 725.

Bibliography

Papers and Articles

Aalto University School of Science. *World record in low Temperatures*. Retrieved July 13, 2016. http://ltl.tkk.fi/wiki/LTL

Alfvén, Hannes. *Quotations*. Cosmology, Retrieved May 27, 2016. http://www-history.mcs.st-andrews.ac.uk/ Quotations/Alfven. html.

Alfvén, Hannes. Cosmology: Myth or Science. Retrieved May 27, 2016..http://alumnus.caltech.edu/~ckank/fringe/alfven/Alfven.html. Alfvén, Hannes. *Hannes Alfven (1908-1995)*.Cosmology. Retrieved June 10, 2016.
http://tmgnow.com/repository/cosmology/alfven.html.

Archer, J. Testosterone aggression: an evaluation of the challenge hypothesis. Neuroisci Bioghav Rev. 2006;30(3):319-45. Epub 2005 Feb 25. Retrieved July 15, 2017.
https://www.ncbi.nlm.nih.gov/pubmed/16483890.

Arp, Halton C. *Additional members of the local group of galaxies and quantized redshifts within the two nearest groups*, in Journal of Astrophysics and Astronomy. September 1987, Volume 8, Issue 3, 241-255. Retrieved July 9, 2016. http://link.springer.com/ article/ 10.1007%2FBF02715046.

Aspect, Alain, Dalibard, Jean and Roger, Gérard. *Experimental Test of Bell's Inequalities Using Time-Varying Analyzers*, Phys. Rev. Lett. 49, 180. Published 20 December 1982. *Retrieved July 13, 2016.* http://journals.aps.org/prl/abstract/10.1103/.

Astronomy news staff. *Universe is Not Expanding After All, Controversial Study Suggests*. May 23, 2014. Retrieved July 17, 2016. http://www.sci-news.com/astronomy/science-universe-not-expanding-01940.html.

Astronomy Magazine. *Galaxies in the early universe mature beyond their years*. Retrieved July 9,2016.
http://www.astronomy.com/news/2014/03/galaxies-in-the-early-universe-mature-beyond-their-years.

Atruio-Barendela, F. *On the Statistical Significance of the Bulk Flow Measured by the PLANCK Satellite*. ArXiv. Retrieved June 25, 2016. http://arxiv.org/abs/ 1303.6614.

Baars, Bernard. A cognitive theory of consciousness.1988, Kindle edition published by the author in 2011. Retrieved Dec. 11, 2016.https://pdfs.semanticscholar.org/20f5/e2242b16acacb8332cc9b834846badce5ce3.pd.f.

Baars, Bernard J and McGovern, Katherine. Global Workspace November 5, 1997. Retrieved July 24,2016. ://cogweb.ucla.edu/CogSci/GWorkspace.html.

Bahcall, John N., Kirhakos, Sofia, Schneider, Donald P. The apparently normal galaxy hosts for two luminous quasars (1995). ArXiv. Retrieved May 30, 2016. http://arxiv.org/abs/astro-ph/9509031.

Barbour, Julian. Killing Time. Uploaded April 8,2008.Retrieved July 1, 2016. https://www.youtube.com/watch?v=WKsNraFxPwk.

Baron d'Holbach. Quotes. Retrieved July 10, 2016. http://.com/www.azquotes author/21200-Baron_d_Holbach.
\
BBC Science and Environment, *Physicists Observe "Negative Mass"*. April 19, 2017. Retrieved April 20, 2017. *http://www.bbc.com/news/science-environment-39642992*.

Bell. On *The Einstein, Podolsky, Rosen Paradox*. November 4, 1964. Retrieved September 4, 2016. https://cds.cern.ch/record/111654/files/vol1p195-200.

Berg, Rob van den. *A Constant that Isn't Constant*. August 9, 2001. Retrieved June 1, 2016. http://physics.aps.org/story/v8/st9.

Bethell, Tom. Rethinking Relativity. Retrieved June 16, 2017. http://www.ldolphin.org/vanFlandern/gravityspeed.html.

Bibliography

Big Bang Cosmythology. *Is Dark Matter Just Plain Hydrogen?* From Sky &Telescope January 2000, page 20. Retrieved May 27, 2016.https://sites. google.com/site/bigbangcosmythology/home/h2.

Blanchet, Luc and Le Tiec, Alexandre. *Model of Dark Matter and Dark Energy Based on Gravitational Polarization*. Physical Review D 78, 024031 Published July 18, 2008. Retrieved May 31, 2016. www2.iap.fr/users/blanchet/images/PhysRevD.78.024031.pdf.

Bode, Stefan; He, Anna Hanxi; Siong Soon, Chun; Trampel, Robert; Robert; and Haynes, John-Dylan. *Tracking the Unconscious Generation of Free Decisions Using Ultra-High Field fMRI*. January 27, 2011.Retrieved July 7,2016. http:// journals. plos.org/plosone/article? id=10.1371/journal.pone.0021612.

Borchardt, Glenn. Quantum Mechanics: A watched pot particle never boils? The Scientific Worldview. April 19, 2017. April 19, 2017. http://thescientificworldview.blogspot.ca/2017/04/quantummechanics-watched-pot-particle.html

Borchardt, Glenn. Do Spent batteries have less mass? The Scientific Worldview. Dec. 7, 2016. Retrieved Dec. 7, 2016. http://thescientificworldview.blogspot.ca/ 2016/12/do-spent-batterieshave-less-mass.html.

Borchardt, Glenn. Elderly Galaxies Plague Big Bang Theory. John Chappell Natural Philosophy Society. Feb.16, 2015. Retrieved April 27, 2011.Retrieved July 7,2016. http:// journals. plos.org/plosone/article?id=10.1371/journal.pone.0021612.

Borchardt, Glenn. Quantum Mechanics: A watched pot particle never boils? The Scientific Worldview. April 19, 2017. Retrieved April 19, 2017. http://thescientificworldview.blogspot.ca/2017/04/ quantummechanics-watched-pot-particle.html.

Borchardt, Glenn. BS for detecting loyalty http://bit.ly/2rMMbs7I The Scientific Worldview. June 7, 2017. Retrieved June 7, 2017.

Borchardt, Glenn. *Do Spent batteries have less mass?* The Scientific Worldview. Dec. 7, 2016. Retrieved Dec. 7, 2016. http://thescientificworldview.blogspot.ca/2016/12/do-spent-batteries-have-less-mass.html.

Borchardt, Glenn. *Elderly Galaxies Plague Big Bang Theory.* John Chappell Natural Philosophy Society. Feb.16, 2015. Retrieved April 3, 2016. http://www.naturalphilosophy.org/site/ glenn.borchardt/ 2015/02/16/elderly-galaxies-plague-big-bang-theory/.

Borchardt, Glenn. *Einstein's most important philosophical* error. Conference of the NPA, 6- 9 July, 2011. Researchgate. Retrieved May 5, 2016. https://www.researchgate.net/publication/ 221706055_Einstein%27s_most_important_philosophical_error.

Borchardt, Glenn. *GPS Does Not Require Relativity.* The Scientific Worldview. October 31, 2012. Retrieved April 20, 2017. http://thescientificworldview.blogspot.ca/2012/10/gps-does-not-require-relativity.html.

Borchardt, Glenn. *Gravitational Attraction is Dead.* April 14, 2016. Retrieved June 15, 2016. http://www.naturalphilosophy.org/ site/glennborchardt/2016/04/14/gravitational-attraction-is-dead/.

Borchardt, Glenn. *Indeterministic propaganda against reality.* The Scientific Worldview. July 27, 2016. Retrieved July 27, 2016. http://thescientificworldview.blogspot.ca/2016/07/. Indeterministic propaganda against reality.html.

Borchardt, Glenn. *Infinite Divisibility of matter and Space.* The Scientific Worldview. March 22, 2017. Retrieved March 22, 2017. http://thescientificworldview.blogspot.ca/2017/03/infinite-divisibility-of-matter-and.html.

Borchardt, Glenn. *Infinite Universe Theory*, Proceedings of the NPA 2007. Retrieved May 28, 2016.

Bibliography

www.scientificphilosophy.com/Downloads/IUT.pdf.

Borchardt, Glenn. *Is Space Matter?* Scientific Worldview. Sept. 30, 2009. Retrieved March 29, 2016. http://thescientificworldview. blogspot. ca/2009/09/is-space-matter. html.

Borchardt, Glenn. *Matter-motion terms in Physics.* The Scientific Worldview. April 20, 2016. Retrieved May 26, 2016. http://thescientificworldview.blogspot.ca/2016/04/ matter-motion-terms-in-physics.html.

Borchardt, Glenn. *Meet George Coyne.* The Scientific Worldview. March 2, 2016. Retrieved March 2, 2016. http://thescientificworldview.blogspot.ca/2016/03/meet-george-coyne.html.

Borchardt, Glenn. *Negative mass?* The Scientific Worldview. May 3, 2017. Retrieved May 3, 2017. http://thescientificworldview.blogspot.ca/2017/05/

Borchardt, Glenn. *Neomechanical Theory of Gravitation.* The Scientific Worldview. December 7, 2011. Retrieved June 22, 2016. www.thescientificworldview.blogspot.ca/ 2011/12/neomechanical-theory-of-gravitation.html.

Borchardt, Glenn. *Resolution of STL-Order Paradox.* Proceedings of the Natural .Philosophy Alliance (2008)v. 5.no.1,p. Retrieved June 20, 2016. www.scientificphilosophy.com/ downloads/ sltorder.pdf.

Borchardt, G. Regressive physics does not know what energy is. The Scientific Worldview. June 28, 2016. Retrieved June 28, 2016. http://thescientificworldview.blogspot.ca/ 2017/06/ regressive-physics-does-not-know-what.html.

Borchardt, Glenn. *The Physical Meaning of $E=mc^2$.* Proceedings of the Natural Philosophy Alliance. January 30, 2016. Retrieved May 27, 2016. http:www.natural philosophy. org/site/ glennborchardt/2015/01/30/11/.

Borchardt, Glen. *The Scientific Worldview and the Demise of Cosmogony.* Progressive Science Institute. Proceedings of the NPA 2007. Retrieved May 3, 2016. www.scientificphilosophy.com/Downloads/TSWATDOC.pdf.

Borchardt, Glenn. *The Ten Assumptions of Science and the Demise of Cosmogony.* Proceedings of the NPA, 2004. Retrieved July 10, 2017. http://www.scientificphilosophy.com/downloads/ttaosatdoc.pdf.

Borchardt, Glenn. *Time is Motion.* Scientific Worldview. The Scientific Worldview. November 30, 20111. Retrieved May 20, 2016. http:/thescientificworldview.blogsopot.ca 2011/11/ time-is-motion.html.

Borchardt, Glenn. *Why time is not an illusion.* The Scientific Worldview Blogspot. December 30, 2015. Retrieved June 12, 2016. http://thescientificworldview.blogspot.ca/2015/12/why-time-is-not-illusion.html.

Borchardt, Glenn and Puetz, Stephen. Albuquerque, New Mexico 2012 Proceedings of the NPA 1. *Neomechanical Gravitational Theory.* Retrieved June 10, 2016.

Broadhurst, Thomas J. Ellis, Richard S. Ellis & Glazebrook, Karl.Faint galaxies - *Evolution and cosmological curvature.* Nature. January 2, 1992. Retrieved May 31, 2016. http://search.proquest.com/openview/368bd3b6f5919d78492fc4d67e9bcd59/1?pq-origsite=gscholar&cbl=40569

Bruno, Michael. *Review of: Gennaro, Rocco J. Higher-Order Theories of Consciousness: An Anthology. Psyche 11 (6),* October 2005. Retrieved December 9, 2016. www.theassc.org/files/assc/2619.pdf.

Bryant, Steven. *Feedback – Responding to Questions and Comments.* August 14, 2016. Retrieved September 2, 2016. http://stevenbbryant.com/2016/08/feedback-responding-to-questions-and-comments/.

Carey, Toni Vogel. Philosophy Now. *Hypotheses (Non) Fingo. 2012.*

Bibliography

Retrieved June 25, 2016. https:// philosophynow.org/ issues/88/ Hypotheses_ Non_Fingo.

CERN. *Dark matter*. 2016. Retrieved May 28, 2016. http://*home.cern/about/physics/dark-matter*.

Chalmers. David J. Facing *Up to the Problem of Consciousness* in the *Journal of Consciousness Studies* 2(3):200-19, 1995. Retrieved June 15, 2016. http://consc net/papers/facing.html.

Chomsky, Noam. Interviewed by Nick day on *Consciousness Central 2017 - Day 2 with guests Noam Chomsky and Hartmut Neven.* June 7, 2017. Retrieved August 15, 2017. Https://www.youtube.com/watch?v=-BgRr53YJ4Q.

Constantin, Anca and Shields, Joseph C. Emission-Line Properties of $z > 4$ Quasars. The Astrophysical Journal, 565:50-62, 2002 January 20. Retrieved December 7, 2016. http://iopscience.iop.org/article/10.1086/324395/fulltext/54203.text.html.

Coyne. George. *Using Mind and Consciousness in Freedom*. August 12, 2015. The Scientific Worldview. Retrieved April 23, 2016. http://thescientificworldview.blogspot.ca/2015/08/using-mind-and-consciousness-in-freedom.html.

Coyne, George. Comment on *Why time is not an illusion, by Glenn Borchardt*. The Scientific Worldview. Retrieved December 30, 2015. Retrieved June 1, 2016. http://thescientificworldview. blogspot .ca/2015/12/ why-timeml.

Coyne, George and Borchardt, Glenn. *Matter and Motion are Abstractions*. The Scientific Worldview. February 3, 2016. Retrieved March 15, 2016. http://thescientific worldview. blogspot.ca/ 2016/02/matter-and-motion-are- abstractions.html.

Coyne, George. *The Myth of "Quantum Entanglement."* The Scientific Worldview. March 23, 2016. Retrieved April 10, 2016. http://thescientificworldview.blogspot.ca2016/03/the-myth-of-quantum-entanglement.html.

Cramer, John G. *The Transactional Interpretation of Quantum Mechanics*. Retrieved July 6, 2016. http://washington.edu/npl/int_rep/tiqm/TI_toc.html.

Csep. *Astronomy 162. Stars, Galaxies, and Cosmology*. Retrieved May 28, 2016. http://csep10.phys.utk.edu/astr162/lect/cosmology/bbproblems.htm.

Csep, *Astronomy 161, The Universal law of Gravitation*. Retrieved May 27, 2016. http://csep10.phys.utk.edu/astr161/lect/history/newtongrav.html.

Davies. P.C.W. *Multiverse Cosmological Models*. 0403047 Australian Centre for Astrobiology, Macquarie University, New South Wales, Australia 2109. Retrieved June, 20, 2016. www.arxiv.org/pdf/astro-ph/.

Dennis, Glenn, de Gosson, Maurice A. and Hiley, B.J. *Bohm's quantum potential as an internal energy. December 17, 2014.*Retrieved June 1, 2016. https://arxiv.org/pdf/1412.5133.

de Hilster, D. *Comparing Aether with the Particle Model*. Proceedings of the CNPS (2017) John Chappell Natrual Philosophy Society.

Dictionary.com. Awareness. Retrieved May 14, 2017. http://www.dictionary.com/browse/awareness.

Dingle, Herbert and O' Keefe, Martin Brian . Science At the Crossroads.1972. Retrieved July, 18, 2017. http://blog.hasslberger.com/Dingle_SCIENCE_at_the_Crossroads.pdf

Dowdye Jr., Edward Henry. *Significant Findings. Gravitational Deflection of Light and microwaves*. 2010. Retrieved July 9, 2016. http://www.extinctionshift.com/SignificantFindings01.htm.

Droege, Paula. *Higher-Order theories of Consciousness*. Internet Encyclopedia of Philosophy. Retrieved December 9, 2016. http://www.iep.utm.edu/consc-hi/.

Bibliography

Djorgovski, Tom S. and Spinrad, H. *Toward the application of a metric size function in galactic evolution and cosmology*. The Astrophysical Journal. December 15, 1981. Retrieved June 18, 2016. http://adsabs.harvard.edu/abs/1981ApJ...251..417D

Einstein, A. *Explanation of the Perihelion Motion of Mercury from GeneralRelativity Theory* 1915. Retrieved July 17, 2017. Gerald Clemence (Rev. Mod. Phys.19, 361364, 1947). http://www.gsjournal.net/old/eeuro/vankov.pdf.

Einstein, A. Geometry and Experience. January 21, 1921. Retrieved June 1, 2016. http://www-history.mcs.st-andrews.ac.uk/Extras/Einstein_geometry.html.

Ellis, Ian. Today in Science. Retrieved December *1, 2016. History.https://todayinsci.com/J/Jeans_James/JeansJames-Quotations.htm.* Einstein, Albert. *Geometry and Experience*. First published 1922. April, 2007. Retrieved July 11, 2016. http://www-groups.dcs.st-and.ac.uk/history/Extras/Einstein_geometry.html.

Evans, Alfred B. quoting Stalin in Soviet Marxism-Leninism: The Decline of an Ideology Stalin, Address to the 16th Congress of the Russian Communist Party (1930), 39. Retrieved November 1, 2017. https://books.google.ca/books?id=ezGGPIze4ZYC&pg=PA39&lpg=PA39&focus=viewport.

Erra, R. Guevara, Mateos, D.M, Wennberg, R. Valazquez, J.L *Towards a statistical mechanics of consciousness maximization of number of connections associated with conscious awareness*. Cornell University Library. Revised January 9, 2017. Retrieved July 25, 2017. https://arxiv.org/abs/160600821.

Famaey, Benoit and McGaugh, Stacey. Modified Newtonian Dynamics (MOND)*: Observational Phenomenology and Relativistic Extension.* May 20, 2012. Retrieved May 31, 2016. ArXiv arxiv.org/pdf/1112.3960.pdf.

Father Vincent. Quoted in Philosophy of Freewill/ Human Functional Freewill. The A3 Society, A New Structure for the New

Millenium.http://a3society.org/HumanFunctionalFreewill.

Feynman, Richard. *Probability and Uncertainty - the Quantum Mechanical View of Nature,* Chapter 6 of *The Character of Physical Law*, 129. Nov 9 to 19, 1964. Retrieved July 10, 2016. http://www. informationphilosopher. com/solutions/scientists/ feynman/ probability_and_uncertainty.html.

Feynman, Richard P. *The Character of Physical Law - 6 -Probability and Uncertainty.* Retrieved July 9,2016. https://www.youtube.com/watch?v =x5RQ3QF9GGI.

Feynman, Richard P. *The Feynman Lectures on Physics.* Chapter 4 *Conservation of Energy, 4-1 What is energy.* Lecture from November. 1964. Retrieved May 26, 2016. http://www.feynman lectures.caltech.edu/I_04.html.

Field, George B. Introduction *The Redshift Controversy.* School of Natural Sciences Institute for Advanced Study. July 13, 2016. http://www.sns.ias.edu/~jnb/Books/Redshift/intro.html.

Flambaum, V.V., Kozlov, Dzuba, M , Angstmann,E., Cheng, Chin, Berengut, Cheng, Chin, J.,Karshenboim, S, Chin, S., Nevsky, A., Porsev, S., Ong, A., Derevianko, A .and others. *Effects of variation of fundamental constants and violation of symmetries* (P, T,) in atoms. Retrieved July 9, 2016. Web2.uwindsor.ca/psas/Abstracts/Flambaum_1.pdf.

Francis, Matthew. *Do We Need to Rewrite General Relativity?* January 18, 2015. Retrieved May 31, 2016. http://www. pbs.org/wgbh/nova/ blogs/physics/2015/06/do-we-need-to-rewrite-general-relativity.

Freedman, Stuart J. and. Clauser, John F. *Experimental Test of Local Hidden-Variable Theories,* Physical Review Letters. April 3, 1972. Retrieved July 10, 2016. http://journals.aps.org/prl/abstract/ 10.1103/PhysRev Lett.28.938.

Fried , J.; Mukamel R.; Kreiman G. *Internally generated preactivation of single neurons in human medial frontal cortex predicts volition.*

Bibliography

February 10, 2011. Retrieved July 7, 2016. klab.tch.harvard.edu/publications/PDFs/gk2630supp.pdf.

Funk, Ken. *What is a Worldview.?* March 21, 2001. Retrieved June 28, 2016. http://web.engr.oregonstate.edu/~funk/Personal/worldview.html.

Gaztanaga, Enrique and Juszkiewicz, Roman. *Gravity's Smoking Gun?*, (2001). April 3, 2001. Retrieved May 11, 2016. http://iopscience.iop.org/article/10.1086/323393/fulltext/015230.text.html

Gastra, Ed. *The Infinite Universe* Retrieved May 30, 2016. http://www.eitgaastra.nl/timesgr/part4/2.html.

Hajdukovic, Dragan Slavkov. *Is dark matter an illusion created by the quantum vacuum?* (2011). June 4, 2011. Retrieved May 31, 2016. http://arxiv.org/abs/1106.0847.

Hajivassiliou, Chris A. *On the cosmological significance of the apparent deficit of small interplanetary scintillation sources.* The SAO?NASA Astrophysics Data System. Retrieved June 15, 2016. http://adsabs.harvard.edu/.

Hansen, Peter M. *Redshift Components of Apparent Quasar-Galaxy Associations: A Parametric Model.* Apeiron. January 2006. Retrieved May 31, 2016. redshift.vif.com/JournalFiles/V13NO1PDF/V13N1HAN.pdf.

Hiley, Basil. E-mails to author. From August 1, 2015 to August 3, 2017.

Hiley, Basil. *From the Heisenberg Picture to Bohm: a New Perspective on Active Information and its relation to Shannon Information.* Retrieved July, 10, 2016. www.bbk.ac.uk/tpru/BasilHiley/Vexjo2001W.pdf.

Hiley, Basil and Callaghan, R E. *Delayed Choice Experiments and the Bohm Approach.* Retrieved July 18, 2016. www.bbk.ac.uk/tpru/BasilHiley/DelayedChoice.pdf.

Howell, Elizabeth. *How Many Stars Are In The Universe?* May 31, 2014. Retrieved May 27, 2016. http://www.space.com/26078-how-many-stars-are-there.html.

Hoyle, Fred. *Quotes by Fred Hoyle.* Retrieved June 1, 2017. http://likesuccess.com/author/fred-hoyle.Hu, E. M., Cowie, L, McMahon,R.J, Capak,P., Iwamuro,F., Kneib, J.P., Maihara, T., Motohara, K. A Redshift z = 6.56. Galaxy Behind the Cluster Abell 370. March 6, 2002. Retrieved May 30, 2016. www.arxiv.org/pdf/astro-ph/0203091.

Hyperphysics. Model of Earlier Events. Retrieved June 21, 2016. http://hyperphysics.phy-astr.gsu.edu/hbase/Astro/ cosmo.html#c5.

IFL Science. *Newly Discovered "Cosmic Wall" Is 1.3 Billion Light-Years Across.* March 13, 2016. Retrieved August 6, 2016. http://www.iflscience.com/space/has-new-cosmic-wall-been-found/.

Internet Encyclopedia of Philosophy. Jerry A. Fodor (1935 —) Retrieved August 7, 2017. http://www.iep.utm.edu/fodor/.

Ijjas, Anna, Steinhardt Paul and Loeb, Abraham. Inflationary paradigm in trouble after Planck. June 25, 2013. *Inflationary schism. Science Direct.* 7 July, 2014. Retrieved June 5, 2016.http://www.science direct.com/science/article/pii/S0370269314004985.

Jain, D. and Dev, A. *Age of High-redshift Objects—A Litmus Test for the Dark Energy Models.* Physical Letters B, Vol. 633, No. 4-5, 2006, pp. 436-440. September 8, 2005. Retrieved May 28, 2016. arxiv.org/pdf/astro-ph/0509212.pdf.

John Chappell Natural Philosophy Society. *Where Critical Thinkers Meet.* Retrieved June 3, 2017. http://www.naturalphilosophy.org/site/about/.

Johnston, Hamish. *Changes spotted in fundamental constant.* Physics World. September 2, 2010. Retrieved June 5,2016. www.physicsworld.com/cws/article/news/2010/sep/02/ changes-spotted-in-fundamental-constant.

Bibliography

Jones, Andrew Zimmerman. *What Is the Copenhagen Interpretation of quantum Mechanics?* 2016. Retrieved May 27,2016. http://physics.about.com/od/quantuminterpretations/fl/What-Is-the-Copenhagen-Interpretation-of-Quantum-Mechanics.htm.

Jones, D.H. et al., *The 6dF Galaxy Survey: final redshift release (DR3) and southern large-scale structures.* March 1, 2009. Retrieved June 28, 2016. http://arxiv.org/abs/0903.5451.

Journal of Astrophysics and Astronomy (1984). *Cosmology: Myth or Science?* http://www-history.mcs.st-andrews. ac.uk/ Quotations/Alfven.html. Dec., 2013. Retrieved May 27, 2016.

Kane, Gordon. *Are virtual particles really constantly popping in and out of existence? Or are they merely a mathematical bookkeeping device for quantum mechanics?* October 9, 2006. Retrieved May 27, 2016. http://www.scientificamerican.com/article/are-virtual- particles-rea/.

Kashlinsky, A. Mysterious Cosmic 'Dark Flow' Tracked Deeper into Universe. March 10, 2010. Retrieved June 20, 2016. http://www.nasa.gov/centers/ goddard/news/releases/2010/10-023.html.

Khamehchi, M.A., Hossain, Khalid, Mossman, M.E., Zhang, Y., Busch, T., Forbes, Michael M. and Engels P. Negative-Mass Hydrodynamics in a Spin-Orbit–Coupled Bose-Einstein Condensate. Phys. Rev. Lett. 118, 155301. Published April 10, 2017. *Retrieved April 20, 2017.* *https://journals.aps.org/prl/abstract/.1103/ PhysRevLett.118.155301.*

Lamoreaux S. K. and Torgerson, R. *Neutron moderation in the Oklonatural reactor and the time variation of α.* March 24, 2004. Retrieved June 10, 2016. https://arxiv.org/pdf/nucl-th/0309048.

LaRocco, Chris and Blair Rothstein. *The Big Bang: It Sure Was Big.* Retrieved May 1, 2016. http://umich.edu/~gs265/bigbang.htm.

LaViolette, Paul .A. *Is the universe really expanding?* Apl 301, 544-553.

January 1986. Retrieved June 18, 2016.
https://www.researchgate.net/ publication/234219969.

Lieu, Richard and Mittaz, Jonathan. *On the Absence of Gravitational Lensing of the Cosmic Microwave Background.* The Astrophysical Journal, Volume 628 no.2. February 17, 2005. Retrieved May 31, 2016. iopscience.iop.org/article/10.1086/429793/pdf.

Lerner, Eric J. *Intergalactic radio absorption and the cobe data.* May.1995. Retrieved May 27, 2016.
adsabs.harvard.edu/full/1995Ap%26SS.227...61L

Lerner, Eric J. *The Big Bang Never Happened.* Retrieved May 27, 2016.
http://bigbangneverhappened.org/.

Lerner, Eric J., Falomo, Renato, and Scarpa, Riccardo. *UV surface brightness of galaxies from the local Universe to z ~ 5.* International Journal of Modern Physics. D, published online May 02, 2014; doi: 10.1142/S0218271814500588. Retrieved July 17,2016.
https://arxiv.org/abs/1405.0275.

Llinás, Rodolfo. *Consciousness and the Brain, Interview with Rodolfo Llinás.* Conducted by Sérgio Strejilevich. Retrieved June 29, 2016.
http://www.cerebromente.org. br/n06/opiniao/llinas_i.html.

Lodato, Simona; Arlotta, Paola (2015-11-13) *Generating Neuronal Diversity in the Mammalian Cerebral Cortex.* Annual Review of Cell and Developmental Biology. *31 (1): 699–720.* November *13* 2015. Retrieved August, 6, 2017. http://www.annualreviews.org/doi/10.1146/annurev-cellbio-100814-125353.

Lui, J. H;; Hansen, D. V; Kriegstein,, A. R. *Development and Evolution of the Human Neocortex. July 8, 2011. Retrieved August 6, 2017.*
https://www.ncbi.nlm.nih.gov/pubmed/21729779.

Lycan, William G. Consciousness and Experience. Retrieved November 5, 2016. https://mitpress.mit.edu/books/consciousness-and-experience.

Bibliography

MacIsaac, Tara. *9 Scientists Who Dispute Theory of Gravity.* The Epoch Times March 30, 2014. Updated June 8, 2016. Retrieved July 10, 2017. http://www.theepochtimes.com/n3/589863-9-scientists-who-dispute-the-theory-of-gravity/full/.

Manor, Peter E. *Gravity, Not Mass Increases with Velocity.* Journal of Modern Physics, 2015, 6, 1407-1411. Published Online August 2015. Retrieved August 7, 2015. https://file.scirp.org/pdf/JMP_2015082810264025.pdf.

Marmet, Louis. *On the Interpretation of Red-Shifts :A Quantitative Comparison of Red-Shift Mechanisms* II (2016) February 29, 2016. Retrieved June 18, 2016. www.marmet.org/cosmology/redshift/mechanisms.pdf.

Marmet, Paul. *A New Non-Doppler Redshift.* Newton Physics. Updated from: Physics Essays Volume. 1, No: 1, p. 24-32, 1988 Retrieved July 09, 2016. http://www.newtonphysics.on.ca/hubble/

Marmet, Paul. *Cosmic Microwave Background Radiation Low Temperature Hydrogen in an Absolute Space as the Likely Cause of the Cosmic Microwave Background Radiation (thus no Big Bang).* Space and Motion. Retrieved May 27, 2016. http://www.spaceandmotion.com/cosmic-microwave-background-radiation.htm.

Marmet, Paul. *Discovery of H_2, in Space Explains Dark Matter and Redshift.* Published in 21st Century Science & Technology, Spring 2000. Last checked February 21, 2016. Retrieved May 27, 2016. http://www.newtonphysics.on.ca/hydrogen/.

Max Planck Institute for Astrophysics. *Is Dark Matter the Source of a Mysterious X-ray Emission Line? April 1, 2016.* Retrieved July 25, 2016. http://www.mpa-garching.mpg.de/332003/hl201604.

Max Planck Institute for Astrophysics. *The diversity of stellar halos in massive disk galaxies.* January 1, 2016. Retrieved September 15, 2016. https://www.mpa-garching.mpg.de/303920/hl201601.

Maxwell, James C. *A Treatise on Electricity and Magnetism/Part*

IV/Chapter XX. Electromagnetic Theory of Light. Retrieved December 9, 2016.
https://en.wikisource.org/wiki/A_Treatise_on_Electricity_and_Magnetism/Part_IV/Chapter_XX.

McGaugh, Stacy. *A possible local counterpart to the excess population of faint blue galaxies,* Nature 367,1994. Retrieved June 18, 2016. www.arxiv.org/abs/astro-ph/9312023.

Mermin, David N. *Is the Moon There when Nobody Looks? Reality and the Quantum Theory.* Physics Today, April 1985, page 38. Retrieved July 14,2016. http://scitation.aip.org/content/aip/magazine/physicstoday/article/38/4/10.1063/1.880.

Merriam-Webster. *Definition of Aware.* Retrieved July 1, 2016. Retrieved May 14, 2017.http://www.merriam-webster.com/dictionary/aware.

Merriam-Webster. *Definition of Cause.* A reason for an action or condition. Retrieved August 14, 2017. https://www.merriam-webster.com/dictionary/cause.

Merriam-Webster. *Definition of Cosmological Constant.* Retrieved April 30, 2016. http://www.merriam-ebster.com/dictionary/cosmological%20constant968.

Merriam-Webster. *Definition of Convergence.* Retrieved May 15, 2016. http://www.merriam-webster.com/dictionary/convergence.

Merriam-Webster. *Definition of Cosmogony.* Retrieved May 16, 2016 http://www.merriam-webster.com/dictionary/cosmogony.

Merriam-Webster. *Definition of Cosmology.* Retrieved April 15, 2016. http://www.merriam-webster.com/dictionary/cosmology.

Merriam Webster. *Definition of Cosmological Constant.* Retrieved April 15, 2016. http://www.merriam-webster.com/dictionary/cosmological%20constant.

Merriam Webster. *Definition of Cycle.* Retrieved September 3, 2016.

Bibliography

http://www.dictionary.com/browse/cycle.

Merriam-Webster. *Definition of Dualism*. Retrieved July 16, 2016. http://www.merriam-webster.com/dictionary/dualism.

Merriam-Webster. *Definition of Immaterialism*. Retrieved April 15, 2016. *http://www.merriam-webster.com/dictionary/immaterialism.*

Merriam-Webster. *Definition of Inertia*. Retrieved May 1, 2016. http://www.merriam-webster.com/dictionary/inertia.

Merriam-Webster. *Definition of Motion. Retrieved May 2, 2016. http://www.merriam-webster.com/dictionary/motion.*

Merriam-Webster. *Definition of Nirvana*. Retrieved December 1, 2016. https://www.merriam-webster.com/dictionary/nirvana.

Merriam-Webster. *Definition of Property*. Retrieved December 13, 2016. https://www.merriam-webster.com/dictionary/property.

Merriam-Webster. *Definition of Self-Awareness*. Retrieved July 1, 2016. http://www.merriam-webster.com/dictionary/self-awareness.

Merriam-Webster. . *Definition of* thing. Retrieved August 14, 2011https://www.merriam-webster.com/dictionary/thing.

Merriam-Webster. *Definition of Time*. Retrieved May 1, 2016. http://www.merriam-webster.com/dictionary/timeMichelson, A.A & Morley, E.W. 1887. *On the relative motion of the earth & the luminiferous ether:* American Journal of Science,v.39, 333-345. Retrieved June 1, 2016. https://en.wikisource.org/ wiki/ On_the _Relative_Motion_of_the_Earth_and_the_Luminiferous_Ether.

Mitchell, Jacqueline. *Insight into the Seat of Human Consciousness*. Beth Israel Deaconess Medical Center November 4, 2016. Retrieved August 18, 2017. http://www.bidmc.org/News/PRLanding Page/2016/November/Fox-Consciousness.aspx.

Moore, G.E. *Determinism and Free Will*. Revised September 23, 2010.

Retrieved July 10, 2016. http://web.mnstate.edu/ gracyk/ courses/web%20publishing/ Determinism_&_Free_Will.htm.

Musser, George. *The Wholeness of Quantum Reality: An Interview with Physicist Basil Hiley.* November 4, 2013. Retrieved May 15, 2016. http://blogs. scientificamerican.com/ critical-opalescence/ the-wholeness-of-quantum-reality-an-interview-with-physicist-basil-hiley/.

NASA, ESA, S. Beckwith (STScl), and The Hubble Heritage Team STScl/AURA. *The Whirlpool Galaxy (M51) and companion galaxy.* April, 25, 2005. Retrieved August 6, 2016. http://hubblesite.org/gallery/album/pr2005012a.

NASA, ESA and the Hubble Heritage Team STScl/AURA Acknowledgment: J. Gallagher (University of Wisconsin), M. Mountain (STScI) and P. Puxley (NSF). *The magnificent starburst galaxy Messier 8.* Released April 24, 2006 Retrieved August 4, 2016. https://www.spacetelescope.org/images/heic0604a/.

Nappi, Bruce. E-mail comments to me on August 5 and 14, 2017 regarding Notfinity Process.

Nave, Rod. *Physical Keys to Cosmology: Difficulties with the Standard Cosmological Model.* Georgia State University. Retrieved June 15, 2016. http:/hyperphysics.phy-astr.gsu.edu/hbase/ Astro/cosmo.html.

Newton, Isaac. *Principles of Philosophy.* Cited in New Mexico Museum of Space History. Retrieved Oct 3, 2016. http://www. nmspacemuseum.org/halloffame/detail.php?id=139.

Nørgaard-Nielsen, Hans U.; Hansen, Leif; Jørgensen, Henning E. Salamanca, Alfonso Aragón; Ellis, Richard S; & Couch, Warrick J. *The discovery of a type Ia supernova at a redshift of 0.31.* June 15, 1989. Retrieved May 15, 2016. http://www.nature.com/nature/ journal/v339/n6225/abs/339523a0.html.

Norsen, Travis and Nelson, Sarah. *Yet Another Snapshot of Foundational Attitudes Toward Quantum Mechanics.* June 20, 2013

Bibliography

Retrieved July 14, 2016. https://arxiv.org/abs/1306.4646.

Odenwald, Sten. *Is space really quantized?* 2011. Retrieved August 3, 2016. http://www.astronomycafe.net/qadir/BackTo286.html.

Odenwald, Sten and Fienberg, Rick. Galaxy *Redshifts Reconsidered.* The Astronomy Café. Sky Publishing Corporation Feb. 1993 Retrieved May 27, 2016.
http://www.astronomycafe.net/cosm/expan.html.

Parvizi Josef and Damasio, Antonio. *Consciousness and the Brainstem.* Cognition, April *2001)79 (1-2): 135–60.* Retrieved June 18, 2016. http://www.ncbi.nlm.nih.gov/pubmed/11164026.

PBS. *People and Discoveries: Big bang theory is introduced* 1927. Retrieved June 29,2016.
http://www.pbs.org/wgbh/aso/databank/entries/dp27bi.html.

Peat, F. David. *Blackfoot Physics: A Journey into the Native American Universe, chapter 1.*
http://www.fdavidpeat.com/bibliography/books/blackfoot1.htm.

Peat, F. David. Gentle Action. Retrieved May 15, 2016.
http://www.fdavidpeat.com/ideas/gentle.htm.

Peat, F. David. *Mathematics and the Language of Nature.* Retrieved May 26, 2016.
http://www.fdavidpeat.com/bibliography/essays/maths.htm.

Peat, F. David. *Language and Linguistics. Bohm and the Rheomode.* Retrieved May 15, 2016.
http://www.fdavidpeat.com/ideas/langling.htm#rheomode.

Peat F. David and Briggs, John. *Interview with David Bohm.* Originally published in Omni, January 1987. Retrieved May 15, 2016.
http://www.fdavidpeat.com/interviews/bohm.htm.

Peat, David. Letter to George Coyne on Bohm's Rheomode. August 12, 201.

Peat, David. Letter to George Coyne on Bohm's ideas about layers of reality.

Percival, Nick. *An Open Letter to the Physics Community The Twin Paradox*. October 2010. Retrieved August 20, 2017. http://worknotes.com/Physics/SpecialRelativity/TwinParadox/htmlpage4.aspx.

Percival, Nick. Twin Paradox. *Data Does Not Match Special Relativity Time Dilation.* June 2015. Retrieved August 20, 2017. http://worknotes.com/Physics/SpecialRelativity/TwinParadox/htmlpage9.aspx.

Physics of the Universe. *The Big Bang and The Big Crunch,* Retrieved May 31, 2016. http://www.physicsoftheuniverse.com/topics_bigbang_antimatter.html.

Plait, Phil. *The Universe is expanding at 73.8 +/-2.4km/sec/megaparsec! So there.* Discover. March 22, 2011. Retrieved September 5, 2016. http://blogs.discovermagazine.com/badastronomy/2011/03/22/the-universe-is-expanding-at-73-8-2-4-kmsecmegaparsec-so-there/.

Progressive Science Institute, *Welcome to Scientific Philosophy.* Retrieved June 1, 2016. http://www.scientificphilosophy.com/.

Qi, Rui. *Quantum mechanics and discontinuous motion of particles.* Retrieved June 1, 2016. https://arxiv.org/pdf/quant-ph/0209022.

Quantum Quotes. *John Wheeler.* Retrieved July 15,290016. http://shorts2015.quantumlah.org/quantum-quotes.

Rense.com. *Big Bang Theory Busted By 33 Top Scientists.* May 27, 2004. Retrieved July 27, 2016. http://rense.com/general53/bbng.htm.

Richter, Phillip; Sembach, Kenneth; Wakker, Bart P. and Savage, D. *Molecular Hydrogen in High-Velocity Clouds.* November 14, 2001. Retrieved May 27,2016. http://iopscience.iop.org/article/10.1086/338050/fulltext/15795.text.html.

Bibliography

Rigler, M.A, & Lilly, S.J. *Infrared surface photometry of 3C 65: stellar evolution and the Tolman signal*, Apl*427* (1994)L79-L82. June 1994. Retrieved June 18,2016. http://adsabs.harvard.edu/abs/1994ApJ...427L..79R

Roach, John. *Unknown "Structures" Tugging at Universe, Study Says Something maybe out there. Way out there*, National Geographic News, November 5, 2008. Retrieved June 20, 2016. http://news.nationalgeographic.com/news/2008/ 11/081105-dark-flow.html.

Rosenthal, David. *Two Concepts of Consciousness*. Article in Philosophical Studies, January. 1986. Retrieved December 10, 2016. https://www.researchgate.net/ publication/ 226411329_Two_Concepts_of_Consciousness.

Sage, Jacob. *Mind, Brain and Consciousness. Do you believe the mind is identical to the brain.*Jan 31, 2011. Retrieved August 1, 2017. ttps://www.psychologytoday.com/blog/mind-brain-and-consciousness/201101/mind-brain-and-consciousness.

Sanders, R. H. *Modified Newtonian dynamics and its Implications.* January 29, 2001. Retrieved June 1, 2016. https://arxiv.org/abs/astro-ph/0106558.

Santilli, Ruggerro. *Nine Theorems of Inconsistency in GRT with Resolutions via Isogravitation.*Galilean Electrodynamics. Summer 2006. Retrieved July 15, 2017. http://www.i-b-r.org/Incons.GravFinalGED-I.pdf. Retrieved July 15, 2017.

Scarpa, Riccardo. M*odified Newtonian Dynamics, an Introductory Review*. January 20, 2006. Retrieved May 31, 2016. www.arxiv.org/pdf/astro-ph/0601478.

Schrödinger, Erwin. *The present situation in quantum mechanics,A Translation of Schrodinger's "Cat Paradox Paper."* Translator: John D Trimmer, Princeton University Press, N.J (1983). Retrieved July 9, 2016. https://www.tuhh.de/rzt/rzt/it/QM/cat.html.

Science Direct. *Electromagnetic radiation*. Retrieved August 14, 2017.

http://www.sciencedirect.com/topics/neuroscience/electromagnetic-radiation.

Science Focus. *What is the largest object in the Universe.? March 5, 2014.* Retrieved July 2, 2016. http://www.sciencefocus.com/qa/what-largest-object-universe.

Science *Quotes. Paul Henri Thiry, Baron D'Holbach.* Today in Science History. Retrieved July 13,2016. http://todayinsci.com/H/Holbach_Paul/HolbachPaul-Quotations.htm.

Science Quotes. Sir James Jeans Quotations. Today in Science History. Retrieved May 27, 2016. http://todayinsci.com/J/Jeans_James/JeansJames-uotations.htm.

Shaw, Duncan. *Aether Concept of Gravity*. John Chappell Natural Philosophy Society 2017 Proceedings. Vancouve, B.C. July 19, 2017.

Shaw,Duncan, W. *Flowing Aether: A concept*. Physics Essays 26, 4 (2013). Retrieved June 26, 2016. www.Duncanshaw.ca.

Shaw. Duncan, W. *Maxwell's Aether: A Solution to Entanglement.* Proceedings of the CNPS (2016). Retrieved June 1, 2016.

Shaw, Duncan W. Outflowing aether. Physics. Essays. Volume 29, Number 4, pp. 485-489.December 2016. Retrieved May 9, 2017. www.Duncanshaw.ca.

Shaw, Duncan W. *Reconsidering Maxwell's Aether*. Physics Essays (27),601(2014). Retrieved June 1, 2016. www.Duncanshaw.ca.

Shaw, Duncan. *The Cause of gravity—a concept*. Physics Essays, 25, 66. May 26, 2012. Retrieved June 1, 2016. www.duncanshaw.ca/11shaw.pdf.

Shaw, Tina. Tech Times. *Our universe is impossible and we shouldn't exist, Higgs-boson scientist says.* June 26, 2014 Retrieved June 4, 2016. http://www.techtimes.com/articles/9131/20140626/our-universe-is-impossible-and-we-shouldnt-exist-higgs-boson-scientist-

Bibliography

says.htm.

Sleator, Daniel. *Quantum Mechanics: Uncertainty, Discontinuity and Interconnectedness. June 6, 1996.* Retrieved June 1, 2016. *http://www.physics.nyu.edu/sokal/transgress_v2/node1.html.*

Smith, Kerri. *Neuroscience vs philosophy: Taking aim at free will* August 31, 2011. Retrieved July 7, 2016. http://www.nature.com/news/2011/110831/full/477023a.htm.

Sokol, Joshua. *Billion-light-year galactic wall may be largest object in Cosmos,* Daily News. March 8, 2016. Retrieved May 31, 2016. https://www.newscientist.com/ article/2079986-billion-light-year-galactic-wall-may-be-largest-object-in-Cosmos/.

Soon, Chun Siong, Brass,Marcel, Heinze, Hans-Jochen, & Haynes, John-Dylan. *Unconscious determinants of free decisions in the human brain.* April 13, 2008. Retrieved July 7, 2016. http://www.nature.com/neuro/journal/v11/n5/abs/nn.2112.html.

Space Daily. *New Look At Microwave Background May Cast Doubts On Big Bang Theory.* April 13, 2008. Retrieved May 31, 2016. http://phys.org/news/2005-08-microwave- background-big-theory.html#n.

Stanford Encyclopedia of Philosophy. *Chance versus Randomness.* August 18, 2010. Retrieved August 8, 2016. http://plato.stanford.edu/entries/chance-randomness/#1.2.

Stanford Encyclopedia of Philosophy. *Quotes Baron d'Holbach.* Retrieved July 13, 2016. http://plato.stanford.edu/entries/holbach/.

Stanford Encyclopedia of Philosophy. *Quotes Baruch Spinoza,* Retrieved July 12, 2016. http://plato.stanford.edu/entries/spinoza/.

Steinberg, Aephraim and associates. *Observing the Average Trajectories of Single Photons in a Two-Slit Interferometer.* University of Toronto's Centre for Quantum Information and Quantum Control. June 2, 2011. Retrieved December 18, 2016.

http://science.sciencemag.org/content/332/6034/1170#login-pane.

Straatman, Caroline. *Galaxies in the Early Universe Mature Beyond Their Years.* Astronomy magazine. March 2014. Retrieved June 3, 2016. http://www.astronomy.com/news/ 2014/ 03/galaxies-in-the-early-universe-mature-beyond-their-years.

Sumner,, William Q. *On the variation of vacuum permittivity in Friedmann universes.* Published July 1994. Retrieved May 29, 2016. http://adsabs.harvard.edu/abs/1994ApJ...429..491S.

Taubes, Gary. *Theorists Nix Distant Antimatter Galaxies.* Science Oct 10, 1997. October 10, 1997. Retrieved July 14, 2016. http://science.sciencemag. org/content/278/5336/226.summary? cited-by=yes&legid=s ci; 278/5336/226.

Techie, Tanya. *Top Ten Scientific Flaws In The Big Bang Theory.* The Tech Reader. *April 26,2015.* Retrieved May 28, 2016. http://thetechreader.com/top-ten/top-ten-scientific-flaws-in-the-big-bang-theory/.

Testcontry.org.10 Hormones That Effect Our Emotions. Retrieved July 15, 2017.https://www.testcountry.org/10-hormones-that-effect-our-emotions.htm?hvid=3qXRwo.

The Physics Hypertextbook. Motion . Retrieved June 18, 2017. https://physics.info/motion/.

The Scientific Worldview website. http://thescientificworldview.blogspot.ca/.

Tryon, Edward P. *Is the Universe a Vacuum Fluctuation?* Nature, vol. 246, p.396–397, (1973). Retrieved May 10, 2016.

University of Alabama. *New Cosmic Look May Cast Doubts On Big Bang Theory.* Retrieved August. 1, 2016. http://www.s8int.com/bigbang6.html.

University of Toronto. *Quantum physics first: Researchers observe*

Bibliography

single photons in two-slit interferometer experiment. June 2, 2011. Retrieved December 18, 2016. http://phys.org/news/2011-06-quantum-physics-photons-two-slit-interferometer.html#nRlv.

Van Flandern, Tom. *Did The Universe Have A Beginning.* Retrieved May 1, 2016. http//metaresearch.org/cosmology/ Did The Universe Have A Beginning.asp.

Van Flandern, Tom. *Quasars: near vs. far.* 1992. Retrieved May 2, 2016. MetaRes.Bull,1, 28-32; http://metaresearch.org.

Van Flandern, Tom. The Speed of Gravity—What the Experoments Say. Physics Letters A.Volume 250, Issues 1–3, 21 December 1998, Pages 1-11. Retrieved June 16, 2017. http:www.science direct.com/science/article/pii/S0375960198006501.

Van Flandern, Tom. *The Top 30 Problems with the Big bang Theory.* Space and Motion. from Meta Research Bulletin 11, 6-13, 2002. Retrieved April 15, 2016. http://www.spaceandmotion.com/cosmology/top-30-problems-big-bang-theory.htm.

Vonk, Jennifer. *Matching based on biological categories in Orangutans (Pongo abelii) and a Gorilla (Gorilla gorilla gorilla).* September 10, 2013. Retrieved May 26, 2016. http://www.ncbi.nlm.nih.gov/pmc/articles/PMC3775627/.

Weisberg, Josh. *The Hard Problem of Consciousness.* Internet Encyclopedia of Philosophy. Retrieved December 19, 2016. http://www.iep.utm.edu/hard-con/.

Wiese, W. and Metzinger, T. Philosophy of Mind and Predictive Processing. July 2017. Retrieved August 6, 2017. http://analyticphilosophy.eu/wp-content/uploads/2017/07/invited_symposium_metzinger.pdf.

West, Michael J., Côté Patrick, Marzke, Ronald O and Jordán, Andres. *Reconstructing galaxy histories from globular clusters.* Nature. January 1, 2004. Retrieved May 28, 2016. http://www.nature.com/nature/journal/v427/n6969/full/ nature02235.html.

Westmiller, Bill. Comment regarding blog *Spooky action at a distance.* The Scientific Worldview. *October 28, 2016.* Retrieved May1, 2016. http://thescientificworldview.blogspot.ca/ 2015/10/spooky-action-at-distanc e.html.

Wikimedia, Commons. *Franco, Francesco (en:User:Lacatosias, User:Stannered) Double slit experiment. May 17, 2011.* Retrieved July 13, 2016. https://en.wikipedia.org/wiki/Double-slit_ experiment#/media/ *File:* Doubleslit.svg via Wikimedia Commons.

Wikimedia. Nekoja, Nekoja. *Double-slit experiment.* December 16, 2005. Retrieved June 3, 2016. https://commons. wikimedia. org/wiki/File:Double-slit.PNG.

Wikimedia Commons. Stigmatella aurantiaca. *Ives-Stillwell Experiment,* July 9, 2012. Retrieved June 1, 2016. https://commons.wikimedia.org/wiki/File:Ives-Stilwell_experiment_DE.svg.

Wikiquote. Richard Feynman from T*he Strange Theory of Light and Matter. Re*trieved May 27, 2916. https://en.wikiquote.org/ wiki/Richard_Feynman.

Wikipedia. *Category Mistake.* Modified June 11, 2016. Retrieved July 19*, 2016.* https://en.wikipedia.org/wiki/Category_mistake.

Wikipedia. *Chunking.* Retrieved July23, 2016. *https://en.wikipedia.org/wiki/Chunking_(psychology.*

Wikipedia. *Damasio's_theory_of_consciousness.* Retrieved *May 15, 2016.* https://en.wikipedia. org/wiki/Damasio%27s_theory_ of_ consciousness..

Wikipedia. *Damped Lyman-alpha system.* Retrieved July 29,2016. https://en.wikipedia.org/wiki/Damped_Lyman-alpha_system.

Wikipedia. *Dark Energy.* Retrieved May 15, 2016. https://en.wikipedia.org/wiki/Dark_energy.

Bibliography

Wikipedia. *Doublethink. Stalin's Address to the 16th Congress of the Russian Communist Party* (1930). Retrieved June 3, 2016. https://en.wikipedia.org/wiki/Doublethink#cite_note-3.

Wikipeida. Field (physics). Retrieved August 14, 2017. https://en.wikipedia.org/wiki/Field (physics).

Wikipedia. *Geometric Transformations*. Retrieved June 15, 2016. https://en.wikipedia.org/wiki/Geometric_transformation.

Wikipedia. *Global Workspace Theory*. Retrieved June 15, 2016. https://en.wikipedia.org/wiki/Global_Workspace_Theory.

Wikipedia. *Metallicity*. Retrieved May 2, 2016, https://en.wikipedia.org/wiki/Metallicity.

Wikipedia. Paul Henri Thiry, Baron d'Holbach from *System of Nature; or, the Laws of the Moral and Physical World* Vol. 1, p. 25. Retrieved June 30,2016. htps://en.wikipedia.org/wiki/Baron_d%27Holbach.

Wikipedia. *Philosophiae Naturalis Principia Mathematica General Scholium*.Third Edition. Retrieved June 25, 2016. https://en.wikipedia.org/wiki/ Philosophi%C3%A6_Naturalis_ Principia_Mathematica.

Wikipedia.*Philosophy_of_Baruch_Spinoza*. Retrieved June15, 2016. https://en.wikipedia.org/wiki/Philosophy_of_Baruch_Spinoza

Wikipedia. *Thalamus* Retrieved July 22, 2016. https://en.wikipedia.org/wiki/Thalamus.

Wikipedia. W*ave propagation.* Retrieved July 27, 2016. https://en.wikipedia.org/wiki/Wave_propagation.

Wikiquote. John Bell. *Free Will*. Quote from edited transcript of radio interview with John Bell (1985). See *The Ghost in the Atom: A Discussion of the Mysteries of Quantum Physics*, by Paul C. W. Davies and Julian R. Brown, 1986/1993, pp. 45-46. Retrieved June 3, 2016. https://en.wikipedia.org/wiki/Free_will.

Zabel, Gary. *Ethics Part One Concerning God* Appendix. May 19, 2003. Retrieved June 15, 2016. http://www.faculty.umb.edu/gary_zabel/Courses/Spinoza/Texts/Spinoza/e1f.htm.

Zyga, Lisa. *Dark matter may be an illusion caused by the quantum vacuum.* August 11, 2011. Retrieved May15. 2016. phys.org/news/2011-08-dark-illusion- quantum-vacuum.htm

Index

A

absolute zero degrees impossible · 174
absolute zero degrees, · 174
abstracta, · 45
abstrere · 44
aether · 30, 31, 41, 50, 72, 73, 74, 75, 90, 93, 97, 98, 104, 129, 132, 133, 134, 166, 167, 169, 174, 180, 183, 185, 187, 188, 201, 205, 212, 220, 221, 222, 223, 224, 225, 229, 230, 231, 232, 233, 236, 237, 238, 239, 240, 244, 245, 246, 254, 260, 261, 262, 263, 264, 265, 315, 317
Aether Model of Gravity · 261
Alfvén, Hannes · 60, 70, 85, 86, 110, 152, 153
alpha
   changes · 167
   decreases found · 169
   increases · 168
antimatter · 105
appearance that galaxies are receding
   due to aether · 74
Aristotle
   time is a measurement of motion · 203
Arp, Halton C. · 31, 33, 71, 72, 93, 94, 97, 99, 108, 109, 119
Arp's QSOs photographs · 99
Atlas of Peculiar Galaxies · 109
Aspect, Alain · 184
attention · 280
awareness · 315

B

Baars, Bernard · 275, 276, 306
Barbour, Julian · 208, 215, 216, 320
Barbour's view of change · 208
Baron d'Holbach · 36, 172, 252, 258, 283
Baryonic matter · 315
BBT · 113
   changing facts to fit theory · 109
   contradicted by voids · 109
   no decrease in space temperature problem · 111
   redshift used as support · 114
   violates 10 assumptions of science · 114
   violates conservation law · 88
   zero energy universe · 66
Bell, John · 61, 184, 185, 192
Bell's theorem · 61, 184, 185
Berkeley, George · 37, 126

BICEP telescope · 189
big bang abstraction · 49
big bang singularity · 84
Big Change · 202
Big Crunch · 202
Big Freeze · 202
Big Rip · 202
blackboard
　Global Workspace Theory · 276
blindsight · 275
blue galaxies excess · 106
body and mind separation confusion · 272
Bohm, David · 37, 40, 51, 60, 177, 180, 185, 186, 187, 200, 202, 203, 210, 211, 213, 225, 230, 250, 251, 252, 253, 254, 255, 256, 257, 258, 259, 263
Bohmian physics · 255, 318
Bohr, Niels · 125, 173
Borchardt · 124
Borchardt, Glenn · 30, 131, 132, 135, 183, 184, 196, 197, 205
　behavior as a univironmental interaction. · 200
　harmful influence of immaterialism · 231
　Heisenberg uncertainty principle · 200
　on Newtonm's non fingo · 243
　on quantum zeno effect · 214
　relativity conflicts with 10 assumptions of science · 160, 162
Bose-Einstein condensate · 134
BOSS Great Wall · 110
brain studies contradict free will
　Haggard, Patrick · 284
　Haynes, John Dylan · 283
　implications · 283
　Roskie, Adina · 285
brightest cluster galaxies · 107
Bryant, Steven · 30, 46, 50, 52, 53, 57, 158, 159, 160, 161, 162, 163, 164, 165, 166, 167, 169, 180, 209, 225, 230, 264
　double-slit experiment explained · 180
　Einstein's significant errors · 161, 162
　problems with Einstein's equations · 161, 163

C

Callaghan, R.J. · 200, 256, 257
Carruthers, Peter · 301
category mistake · 272
Causality · 37, 250, 258, 259
Causality and Chance in Modern Physics · 250, 258
Chabris, Christopher · 280
Chalmers, David · 293
change the facts to fit BBT · 107
chaos and heat death · 70
Chapman, Sydney · 60, 86
Chomsky, Noam · 45, 305
chopped vegetables · 294
chunking process · 277
CMB
　alternate explanations · 114, 115
cognition · 44, 46, 49, 52, 297
Collingwood, Robin · 36
complementarity · 38, 127
complexification · 239, 244, 315
compression · 46, 48, 52, 224
Compton effect · 223
computational theory of the brain and mind · 304, 312
Conscious Access hypothesis · 277
consciousness · 31, 269, 294
　Blindsight · 270

Index

hard problem · 31
consciousness. · 31, 272, 275, 277, 278, 279, 280, 281, 284, 289, 290, 291, 293, 294, 295, 296, 299, 302, 303, 308
conservation · 36, 38, 66, 89, 90, 92, 98, 186, 197, 230, 253
consupponible · 36, 315
contradictory densities problem · 101
convergence, · 246, 264
convergence. · 315
Copenhagen interpretation · 58, 61, 126, 127, 181, 209, 251, 315
Cornille, Patrick · 223
cosmogony · 33, 66, 67, 85, 90, 98, 209, 230
cosmological constant · 79, 100, 107, 113, 316
cosmological principle · 67, 68, 109, 116
cosmological redshift, · 93
cosmos age problem with globular clusters · 94
creatio ex nihilo · 70, 85
curved space · 70, 160, 229
cycles · 53, 203, 235

D

Damasio, Antonio · 288, 289, 290
dark energy · 48, 83, 91, 98, 104, 105, 107, 220, 224, 235, 316
dark flow · 102
dark matter · 48, 77, 83, 97, 100, 103, 104, 111, 114, 219, 220, 224, 235
de Broglie, Louis · 251
Delayed-Choice Experiments and the Bohm Approa · *See* Hiley and Callaghan
Delayed-Choice Experiments and the Bohm Approach · 200
Descartes, René · 271, 272, 275
Descartes,René
  Cogito ergo sum · 271
determinism · 34, 36, 37, 184, 185, 215, 232, 252, 253, 254, 255, 283, 285, 311, 316
  hard · 283
discrete types · 52
Dispositional HOT Theory · 301
Disruptive · 46, 52, 159, 160, 162, 180, 209, 210, 225, 229, 264
Doppler effect · 71, 72, 75, 82, 84, 93, 94, 104, 165, 320
double-slit experiment · 61, 134, 175, 177, 180, 188, 225
Dowdye, Edward · 160
Dowdye, Edward · 160

E

$E=mc^2$ · 131, 132, 133, 154, 160, 162, 234, 248
easy problem of consciousness
  functional evauation · 292
Einstein
  on Geometry and Experience · 59
Einstein on math · 56
Einstein, Albert · 18, 30, 35, 46, 50, 53, 56, 67, 70, 79, 107, 114, 115, 134, 135, 158, 159, 160, 161, 162, 163, 165, 182, 183, 184, 190, 203, 204, 209, 223, 224, 231, 236, 246
Einstein's Errors · 46, 53
Einstein's Errors · 162
Einstein's types error · 53
electromotive force · 188
elliptical galaxies · 110

energy · 198
- as a matter-motion calculation · 31, 131, 133, 199
- as a thing abstraction · 130, 133
- as defined by orthodox physicists · 198

Engels, Peter · 134
entanglement · 30, 186, 188, 220
entropy · 38, 70, 78, 101
EPR, · 183
equivalence principle · 70, 167, 169, 316
Euclidean space · 84
Evolutionary Benefit of Advanced Mental Processes · 297
existence status · 46, 50, 52
expansion of Universe · 31, 33, 49, 50, 67, 68, 71, 74, 76, 78, 79, 80, 81, 82, 83, 84, 86, 87, 96, 97, 102, 104, 106, 107, 108, 111, 113, 116, 229, 316, 320
expansion tests · 79, 80, 81, 84, 221

F

Faraday effect · 108, Faraday rotation
Feynman · 32, 48, 57, 128, 181, 190, 224, 264
fine structure constant · 167, *alpha*
Flambaum, Victor . 167, 168
flatness" or "oldness" problem · 87
Fodor, Jerry · 312
fourth assumption of science · 132, 295
Fox, Michael D. · 268
free will · 281, 283, 284, 285, 312, 317
Frequency · 53
Friedmann, Alexander · 79

G

Galaxy Evolution Explorer · 83, 108
Galilei, Galileo · 56
Gamow, George · 98, 231
Gazzaniga, Michael · 285
General Relativity Theory · 31, 48, 51, 54, 70, 73, 125, 135, 160, 162, 167, 169, 205, 225, 235, 260, 317
gentle action · 212
Glannon, Walter · 285
Global Workspace Theory · 275, 276
grand unified theory · 31
gravitational energy · 91
Gravitational Pressure Gradient · 237, 240, 246
gravitational shadowing · 231
gravity, defined as a distortion of space, · 135
Greene, Brian · 78
GRT. · 51, 54, 135, 260, 317
Guth, Alan · 76, 77, 87, 89, 120, 327

H

Hajdukovic, Dragan Slavkov · 103
Hansen, Peter · 109, 120
happiness · 45, 291
hard problem of consciousness · 276, 293, 294, 295, 297
HOR solution · 300
Haynes, John Dylan · 283, 284
Heisenberg, Werner · 37, 61, 69, 125, 126, 127, 128, 153, 200, 215, 250, 258, 318
Higgs particle · 189
Higher-order Perception Theory · 302
Higher-order Thought Theory

Index

characteristtics · 300
Hiley and Callaghan
   Delayed-Choice Experiments and the Bohm Approach · 257
Hiley, Basil · 186, 187, 200, 254, 255, 256, 257, 258, 259
   Two Letters · 178
holomovement · 255
HOR/HOT theories · 297, 298, 299, 300
horizon or isotropy problem · 86
horizon problem · 78, 87
Hoyle,, Fred · 33
Hubble Ultra Deep Field · 112
Hubble, Edwin · 49, 71, 72, 104, 107, 108, 112, 113, 115
Hubble's law · 49

I

idealism · 62, 126, 134, 232
Ierome Bernard Cohen · 243
immaterialism · 126, 231, 273, 315
infinite universe · 97, 189, 196, 201, 205, 214, 231
Infinite Universe Theory · 30, 60, 68, 116, 122, 125, 229, 230, 231, 232, 233, 234, 247, 254, 316, 321
infinity · 36
Infinity · 230
inflation · 68, 69, 76, 77, 78, 86, 87, 89, 90, 91, 98, 104, 105, 116
inflation's main problems
   multiverse, Lioiville and entropy · 78
Inflationary theory criticism · 77
   Steinhardt, Paul · 78, 87
inseparability · 36, 85, 93, 230, 252, 254, 295
instantiated · 45, 46, 47, 49, 204, 206, 317
instantiation · 46, 47, 50, 51, 61, 209, 317
interconnection · 36, 40
irreversibility · 39
Ives, Herbert · 165, 221
Ives-Stillwell · 165

J

Jackson, J.D> · 58
Jeans, James · 56

K

Kashlinsky, Alexander · 101, 102
kinetic energy, · 132, 224

L

Lamb shift · 69
language of math · 58
Lemaître, Georges · 70, 98
Lerner, Eric J. · 82, 83, 101, 120, 124
Levin, Daniel · 280
Lieu · 114, 115
Lieu, Richard and Mittaz, Jonathan · 114
Lightman, Alan · 174
Llinás, Rodolfo · 278, 280

M

Marmet, Louis · 74
massless photons myth · 134
material process · 46, 49, 52
materialism · 37, 252, 319

361

math limitations in theories · 59
Mathematicians do abstractions · 54
matter · 173
matter is an abstraction · 195, 196
matterless motion · 129
matterless object · 125, 133
matter-space continuum · 196
mature BBT
    problem of mature galaxies 12 billion light years away · 112
Max PlanckInstitute for Asrophysics (MPA(· 19
mega-vortex · 102, 319
Mele, Al · 284, 285
Mermin, David · 128
Methodological motivation · 298
Michelson and Morley experiment · 166, 167, 264
microcosms · 173
Milgrom, Mordehai · 220
Minkowski spacetime · 50, 73
Modern Mechanics · 30, 162, 164, 166, 190, 210, 225
    exceeds relativity's accuracy · 162, 164
    geometric transformation. · 164
    motion explained · 164
    objects faster than speed of light · 169
molecular hydrogen · 72, 104, 129, 230
    accounts for redshift and dark matter · 72
    provides heat of CMB · 75
MOND · 100, 103, 111, 220
Moore
    G.E. · 283
motion · 206
motionless matter absurd idea · 174
Mysterious Universe · 56
Myth of Quantum Entanglement · 192

N

nadal · 30, 53, 164, 206, 208, 210, 216
Nadal · 208, 319
Nave, Rod · 88, 105
n-body simulations · 99
negative mass myth · 134, 135
Neomechanical Gravitational Theory · 41, 233, 235, 236, 237, 238, 240, 244, 245, 246, 260
neurologists · 268, 274
neuroscience discoveries · 298
neuroscientists · 268, 286, 293, 296, 298
Newton, Isaac · 30, 35, 57, 69, 129, 135, 205, 209, 221, 231, 240, 241, 243, 248, 311
Newton's law of universal gravitation · 69
notfinity · 202, 205, 313
Notfinity · 203
nuclear fission · 133

O

Observer illusion · 281, 285, 286, 287, 325, 329
Our Mathematical Universe · 57

P

particles exceed predicted maximum energies · 111
particle-wave duality · 180
Patelke, Elizabeth · 252
Peat, F. David · 60, 202, 203, 210,

Index

211, 212, 213
letter to me on infinite layers · 203
Peat's letter on Bohm's rheomode · 211
Phenomenological motivation · 298
photoelectric effe · 223
physicality · 46, 51, 52
pilot wave theory · 225, 251
Planck · 40, 75, 78, 101, 167, 183, 212, 214, 219
Plasma Cosmology · 98, 106
Popper, Karl · 68
principle of Equivalence · 231
probability waves · 57, 178, 223
programmers use abstractions · 54
Propositions about predicates · 54
protoself · 289, 290
Puetz, Stephen · 30, 74, 102, 119, 131, 167, 209, 233, 234, 235, 237, 238, 245, 246, 247, 248, 316, 323

Q

QMT · 31, 32, 34, 48, 51, 57, 58, 60, 61, 62, 90, 91, 125, 126, 128, 133, 134, 172, 173, 174, 175, 176, 177, 178, 180, 181, 182, 184, 186, 187, 188, 189, 190, 194, 210, 213, 214, 215, 220, 225, 226, 229, 250, 251, 253, 254, 255, 258, 315, 320, 321, 322
qualia · 293, 296
quantum entanglement · 182, 183, 185, 186, 187, 188
Quantum Field Theory · 54
quantum mechanics theory · 30, 48
quantum potential · 185, 186, 255, 256, 257, 258
quasar's intrinsic redshifts · 74

R

Ratey, John · 280, 281, 307
realism · 62, 134, 184, 232
redshift · 19, 71, 72, 74, 93, 97, 106, 119, 320
  Aethereal Theory · 75
  is intrinsic · 33
  non-expansion mechanisms · 74
  quantization · 74, 84, 99
referencing · 46, 48, 52, 67
relativism · 40, 197, 237, 238, 245, 253
relativity physics · 125, 133, 167, 194, 225
relativity theory · 30, 48, 51, 58, 60, 62, 104, 125, 158, 159, 162, 164, 165, 167, 175, 210, 225, 236
rheomode · 210
Rosenthal, David · 298, 300
Rosenthal, David M. · 300
Ryle, Gilbert · 272

S

Sagnac, Georges · 166, 263
Schrödinger equation · 127
Schrödinger, Erwin · 127, 175, 178, 181
Schrödinger's cat. · 181
Sciama, Dennis · 89
second law of thermodynamics · 38, 70, 71, 101
self-awareness · 298
separability · 133
Shaw, Duncan · 30, 183, 187, 188, 220, 221, 222, 223, 224, 240, 260, 261, 262, 263, 264
Shaw,Duncan · 62, 226, 266

Simons, Daniel · 280
simplifying · 46, 51, 52
space as aetherl · 133
space without matter · 129, 245
spacetime · 48, 50, 52, 54, 60, 66, 67, 70, 73, 125, 132, 134, 159, 165, 185, 202, 205, 206, 207, 212, 225, 229, 236, 317
spacetime continuum · 50, 125
Special Relativity Theory · 205
specialist cortical processing regions, · 298
speed of light · 69, 77, 86, 159, 166, 167, 169, 238, 260, 321
spherical wave proof · 53, 158, 159, 160, 162
Spinoza, Baruch · 34
SRT · 31, 48, 51, 125, 131, 160, 162, 165, 166, 167, 169, 220, 225, 320
Stalin, Joseph · 172
State consciousness · 299
Steinberg, Aephram · 251
Stillwell experiment · 166
Stillwell, G.R. · 165, 166
Straatman, Caroline · 112, 124
superposition · 51, 175, 178, 184, 185, 186
surface brightness · 80, 81, 82, 83, 108, 120
systems philosophy · 35

T

Taubes, Gary · 105, 123, 352
Tegmark, Max · 57, 60
thalamocortical system · 278
*The Emperor's New Clo*thes · 190
The Ten Assumptions of Science · 35, 36, 41, 160, 162, 170, 209, 230, 237

The Undivided Universe · 254, 256, 259
theory of everything · 31, 57
thing
   as a thing abstraction · 131
Through the Looking Glass · 78
time · 52, 70, 207
   as a verb · 204, 206
   as motion · 30, 39, 50, 52, 99, 125, 160, 182, 196, 203, 204, 205, 206, 208, 231, 319, 321
   change reefers to result of motion · 209
   is motion · 73
   is not an illusion · 205
   universal · 207
time dilation absurd abstraction · 165, 166
time equals moving · 207
time is motion · 30, 39, 73, 196, 206, 215
time is not an illusion · 205
timing · 203, 204, 205, 206, 207, 216, 284, 321
tired light hypothesis · 84
total-mass equation · 244
Tryon, Edward · 89, 90, 91
types · 46, 50, 52, 53, 117, 135, 160, 162, 197, 210, 278, 279, 291
   compound · 46, 52, 53
   discrete · 46, 52, 53, 90, 166, 224

U

uncertainty · 37, 200, 215
uncertainty principle · 69, 127, 200, 250, 315, 318, 320
Universal Cycle Theory · 30, 131, 209, 235, 246, 248, 254, 323, 325
universals · 54
Universe

Index

definition · 321
my abstraction · 28
univironment · 35, 322
univironmental determinism · 29, 35, 184, 322

V

vacuum energy · 68, 69, 77, 89, 107, 316
vacuum fluctuation. · 90
Velocity · 53
Venn diagrams and set theory · 61
Venturi effect · 262
virtual particles · 50, 69, 91, 118
void · 31, 50, 67, 68, 71, 77, 90, 91, 225, 229, 230, 232, 233, 236
von Leibnitz, Gottfried Wilhelm · 271
Vonk, Jennifer · 44
vortex · 235, 237, 239, 245, 246, 317

W

Walker, Friedman Robertson · 108
wave packet" or "wave train · 73
Wavelength · 53
wave-particle · 62, 93, 127, 173, 174, 177, 222, 254
wave-particle duality illusion · 61, 134, 318
Webb, John · 168
Weinberg, Steven · 200
Weizsäcker,, Carl Friedrich · 126
weltanschauung · 34
Westmiller · 192
Westmiller, Bill · 183
Wheeler, john · 181, 256, 257, 258
Wheeler, John · 256
Wheeler's delayed choice experiment · 256
Wide-Intrinsicality View · 300
Wigner, Eugene · 56
working memory · 275, 277, 278, 279, 281, 318

Z

Zeno Effect · 214
Zeno's arrow paradox · 214
zero-energy universe · 89, 90, 91, 98
zero-point energy · 50, 92, 174, 216

CPSIA information can be obtained
at www.ICGtesting.com
Printed in the USA
LVOW12s1356230917
549812LV00002B/357/P